What Genes *Can't* Do

Basic Bioethics
Glenn McGee and Arthur Caplan, series editors

What Genes *Can't* Do

Lenny Moss

A Bradford Book
The MIT Press
Cambridge, Massachusetts
London, England

First MIT Press paperback edition, 2004

© 2003 Massachusetts Institute of Technology

This book was set in Sabon by SNP Best-set Typesetter Ltd., Hong Kong and was printed and bound in the United States of America.

Library of Congress Cataloging-in-Publication Data

Moss, Lenny.
  What genes can't do / Lenny Moss.
    p.   cm.—(Basic bioethics)
  "A Bradford Book"
  Includes bibliographical references and index.
  ISBN 0-262-13411-X (hc : alk. paper), 0-262-63297-7 (pb)
  1. Genetics—Philosophy. I. Title. II. Series.

QH430 .M674   2002
660.6′5—dc21
                                 2001056298

10 9 8 7 6 5 4 3 2

To my daughters, Julie and Rachel

Contents

Series Foreword

We are pleased to present the sixth volume in the series Basic Bioethics. The series presents innovative book-length manuscripts in bioethics to a broad audience and introduces seminal scholarly manuscripts, state-of-the-art reference works, and textbooks. Such broad areas as the philosophy of medicine, advancing genetics and biotechnology, end-of-life care, health and social policy, and the empirical study of biomedical life will be engaged.

Glenn McGee
Arthur Caplan

Basic Bioethics Series Editorial Board
Tod S. Chambers
Carl Elliot
Susan Dorr Goold
Mark Kuczewski
Herman Saatkamp

Acknowledgments

The following individuals contributed to the realization of this work and have my everlasting gratitude: Mina Bissell, Paddy Blanchette, Jeff Botkin, Nancy Burke, Candace Brower, Andrew Cooper, Caroline Damsky, Mary Davis, Alexis Kurland Deeds, Eric Kurland Deeds, Bert Dreyfus, Brian Eden, Arthur Fine, Paul Griffiths, Joseph Heath, David Hull, Kuni Kaneko, Karin Klein, Elizabeth Lloyd, Ed Manier, Thomas McCarthy, Paul Millea, Leslie Mulligan, John Opitz, Susan Oyama, Beverly Packard, Gordon Parry, Bill Ramsey, Bob Richards, Kate Ryan, Phil Sloan, John Stubbs, Suzanne Tronier, Steve Watson, Steve Weinstein, and Bill Wimsatt. Extra-special thanks go to Angela Alston, Carolyn Gray Anderson, David Depew, Steve Downes, Sabrina Haake, Glenn McGee, Bob Perlman, and Harry Rubin. A portion of this research was supported by funding from the ELSI program through the Center for Genetic Science in Society at the University of Utah.

Introduction

There can be little doubt that the idea of "the gene" has been the central organizing theme of twentieth century biology. And biology, especially since the inception of its molecular revolution of the 1940s and 1950s, has become increasingly influential in academic venues, including philosophy, public life and policy, medicine and the health sciences generally, and in everyday self-understanding. More recently the promises and prospects of new biotechnology to cure diseases and offer novel reproductive options, and correlatively to win or lose fortunes on the stock market, both further magnified by media attention associated with the Human Genome Project, have brought the gene into even greater public prominence. The selection of my topic was motivated by the high stakes that our understanding of biology has come to have on a variety of levels. Intellectually, in its impact on the humanities, arts, and human sciences; ethically, in its formative effect on human identities and our underlying interpretation of what it means to be human; and socially, with respect to the defining, normalizing, and pathologizing of human difference. The title *What Genes Can't Do* is meant to recall *What Computers Can't Do* by Hubert Dreyfus and to suggest, by analogy, my aspiration to influence a powerful social-technical trend by way of a philosophically guided, empirically argued critique.

Philosophically speaking, what I hope to provide is a platform upon which new naturalistic lines of thought can interweave biological and sociocultural threads at a very fundamental level. Above all, the present work hopes to contribute to freeing the "naturalizing" enterprise from the unnecessary burdens of preformationistic baggage and thereby better to allow for the re-embedding of the self-understanding of human language and knowledge in contingent social and developmental processes.

In the course of attempting to get the story right about genes a wide variety of issues are addressed, often in the context of minidialogues with a variety of key contributors, some from the remote past and others from the hot-off-the-press present. In order to help the reader ward off the danger of losing the forest for the trees, I will first offer a brief overview of the text.

The work begins with a wide-ranging historical reconstruction and conceptual analysis of the meaning of "the gene" that results in defining and distinguishing two different genes. Each of these can be seen as an heir to one of the two major historical trends in explaining the source of biological order: preformationism and epigenesis. The preformationist gene (Gene-P) predicts phenotypes but only on an instrumental basis where immediate medical and/or economic benefits can be had. The gene of epigenesis (Gene-D), by contrast, is a developmental resource that provides possible *templates* for RNA and protein synthesis but has in itself no determinate relationship to organismal phenotypes. The seemingly prevalent idea that genes constitute information for traits (and blueprints for organisms) is based, I argue, on an unwarranted conflation of these two meanings which is, in effect, held together by rhetorical glue. Beyond this historical, conceptual, and rhetorical inquiry the bulk of this work then concerns itself with an empirically up-to-date analysis of the cell and molecular basis of biological order and of the pathological loss of the same.

In each of these chapters I structure my analysis with the idea in mind that the conflated view can be held empirically accountable. I do not conjure up straw men to represent that position but rather use what I take as the most historically influential formulations of the gene-as-information-for-phenotypes position as my points of reference. In the first instance I use Schrödinger's early and highly influential thermodynamic argument for why a solid-state "aperiodic crystal" *must be* the core of biological order.

The task here is really twofold: first to recall that Schrödinger had a real argument for the promotion of his famous hereditary code-script metaphor and second to indicate how and why we can now see that his argument was mistaken. In the second instance, that concerning disorder, my principal point of reference is the somatic mutation hypothesis. Albeit in evolving forms, some version of the somatic mutation hypothesis has dominated cancer biology throughout the twentieth century. I mean to show how a bifurcation in the understanding of cancer commenced with the "phylogenetic turn" that took place at the beginning of the twentieth century and resulted in an ongoing dialectic between genetic determinists at the center and developmentalists at the margins. I have reconstructed the history and fortunes of the somatic mutation hypothesis program in part to give shape and philosophical meaning to the recurrent challenges that have been brought forth from the margins.

Philosophers of biology have hitherto steadfastly avoided the topic of cancer (almost as if talking about it could make it catching). But in considering the loss of organismic order (and the corresponding emergence of malignant order) philosophically charged questions about the distinction between normal and pathological come quickly to the fore. Carefully considered, the research trajectory of the somatic-mutation hypotheses when confronted by its own empirical shortcomings provides some of the most cogent evidence *against* the conflated preformationist view from which it arose.

With this somewhat bare-boned structure in mind, I will now try better to prepare the reader for some of the winding curves and vistas that come up along the way. A main objective of chapter 1 is to account for how a putatively misguided notion of the gene could have possibly arisen and in so doing to clarify just what is conceptually at issue. My principal strategy is that of reconstructing the conceptual pathway to our contemporary genes as a highly contingent transformation of those basic life concepts which held sway during the nineteenth century. Telling this story is complicated by the need to debunk two pervasive myths about the life sciences—namely, that *real* biology only begins with Darwin and that the conceptual ground of genetics owes its existence to some chancy rediscovery of the work of Mendel.

The particular bone I have to pick with these myths has nothing to do with the giving or taking of scientific credits but rather with their role as impediments to a coherent conceptual history of our most basic biological concepts. With respect to the nineteenth century, I have benefited especially from Lenoir's analysis of the role of a distinctively neo-Kantian teleological heuristic in guiding the central stream of early nineteenth century morphology and physiology. Where I have taken some initiative is in analyzing the conceptual path whereby the holistic notion of Kant and Blumenbach, that of a 'stock of *Keime und Anlagen*, becomes reformulated, under the pressure to accommodate Darwinian processes of variation and selection, into an agglomeration of parts. It is in this transition, and in large measure as a kind of conceptual side effect, that the holistic potential and thus the adaptive agency of the living organism was lost to the invisible hand of natural selection and, further downstream, to "selfish" genes. My claim would be that one simply cannot appreciate the twists and turns and the tensions and bifurcations of twentieth century biology without recognizing the stakes that were set up by that transition. I have referred to this transition as the "phylogenetic turn" to mark the movement away from ontogeny and toward phylogeny as the new center of gravity for the explanation of biological form.

The conceptual *chunk-of-anlagen* that Mendel dubbed the unit-character did indeed become the prototype for a new genetic preformationism, but, as Raphael Falk has argued, for Mendel it was only meant to serve as an instrumental function for breeders and not as a universal theory for all of biology. The path from an instrumental to a constitutive attribution of status to the chunk-of-anlagen is recounted, as is the suppression-marginalization of the hereditary role of the cytoplasm (with thanks to Jan Sapp).

Special emphasis is placed on the insightful and critical reflections of Wilhelm Johannsen. It was after all Johannsen who introduced the terms "gene," "genotype," and "phenotype" and who did so precisely *as a critique* of preformationist fallacies and on behalf of a return to a holism defined in terms of the full range of developmental phenotypic potentials associated with any genotype. Several pages are devoted to considering the real contemporary relevance of Johannsen's stunning reflections of 1926. Using Johannsen as a point of departure, I introduce my distinction between the preformationist Gene-P and the epigenesis Gene-D, and from this follows a consideration of what it would mean for a gene to satisfy the conditions for being both a Gene-P and a Gene-D simultaneously.

The empirical fruits of several decades of research in molecular, cell, and developmental biology have revealed that what distinguishes one biological form from another is seldom, if ever, the presence or absence of a certain genetic template but rather *when* and *where* genes are expressed, *how* they are modified, and into *what* structural and dynamic relationships their "products" become embedded. If genes are to be both molecules which function as physical templates for the synthesis of other molecules *and* determinants of organismic traits and phenotypes, then somehow genes would have to, in effect, provide their own instructions for use. They would have to be able to specify when and where their templates would be put to use, how such products would be modified and targeted, as well as in what structural and dynamic relationship they would reside. Indeed, it is just this sense of genes being able to do this which appears to be conveyed with references to genes as information, as programs, as blueprints, as encyclopedias of life, and the like.

Following the strategy of chapter 1, chapter 2 examines the historical genesis of the genes-as-text metaphor, but in so doing a new set of issues arises. The growth of the gene-as-text discussion appears to veer off from empirical reality (or perhaps becomes central to determining what would count as empirical reality). The idiom of the language-of-the-gene became written not by those whose hypotheses were successful but rather by those whose metaphors were successful. The intuitions of postmodernist critics who see a runaway rhetoric of life that simply constructs itself cannot be responsibly ignored. Chapter 2 embarks on several lines of inquiry. In locating the foundation of the textual gene-talk in Schrödinger's notion of the hereditary code-script, I show that what becomes a rhetorical tradition begins with an interesting, *empirically accountable*, argument which has since been forgotten or ignored, and I go to some length to explicate that argument. The subsequent divergence of empirical and rhetorical achievements is adumbrated by the benefit of the recent work of Lily Kay.

I then explore the cognitive consequences of this new rhetoric of life in the course of examining the perceptive insights of rhetoric-of-science critic Richard Doyle. Ultimately, however, I take issue with Doyle over what appears to be his own tacit methodological complicity with that *utonomization of rhetoric* that he ostensibly means to be criticizing. With the interpretive sensitivity of a good literary critic, Doyle exposes the semantic stakes in a manner that far outreaches any narrowly analytical talk about intertheoretic reductionism or the like. However, when it comes to the practical-normative dimension of social-intellectual critique, Doyle simply drops the ball.

Chapter 3 is principally concerned with clarifying the cellular and molecular basis of biological order using Schrödinger as a point of departure. It is precisely in light of the semantic consequences of the conflated gene-rhetoric (and all the ramifications of this suggested above) that the basis of such rhetoric—to the extent that there is one—cannot be left unexamined. Schrödinger argued that only the thermodynamics of the solid state (and thus the "aperiodic crystal"), newly (for him) revealed by quantum mechanics, could account for the existence and continuity of biological order. Whether the subsequent history of empirical investigations have ruled in his favor or not must be made relevant to the force of his rhetorical legacy. Ongoing claims on behalf of what I refer to as the "conflated gene" must be held empirically accountable.

The principal intention of chapter 3 is to demonstrate that biological order is distributed over several parallel and mutually dependent systems such that no one system, and certainly no one molecule, could reasonably be accorded the status of being a program, blueprint, set of instructions, and so forth, for the remainder. The idea of characterizing three subcellular epigenetic systems is derived from Jablonka and Lamb, although I signficantly depart from them in my treatment of the first two of these systems (organizational structure and steady-state dynamics). With respect to the former, I offer a fairly detailed account of the differentiated, membrane-based, structural, and functional compartmentalization of the cell. Biochemically distinct membranous bodies constitute the necessary and irreplaceable templates of their own production and reproduction, are passed along from one generation to the next, and provide the unavoidable context in which DNA can be adequately interpreted, that is, in which *genes can be genes*.

Under the heading of steady-state dynamics, I offer an extended discussion of the theoretical work of Stuart Kauffman. Kauffman's work is most relevant here because he too presents an explicit response to

Schrödinger, relying, in his case, on the implications of computer-simulated models of complex nonlinear systems. Kauffman finds that given certain internal parameters a complex system will gain "order for free" by converging on some comparatively small number of attractor states. Kauffman's model, while complex as a formal system, is still far simpler than any actual biological system. While his work provides a powerful window into the nonintuitive, self-ordering consequences of the dynamics of nonequilibrium complex systems and an important rejoinder to Schrödinger and his descendents, it is shown to pose reductionistic dangers of its own.

The final category of parallel epigenetic systems that I consider is that of chromatin marking. My discussion here is comparatively brief and pertains to the ability of cells to chemically modify genomic DNA in developmentally and environmentally sensitive ways, including the gender-specific chromosome marking found in all mammals and referred to as *imprinting*. The larger significance of such mechanisms may only come to be fully appreciated when the extent of the plasticity of DNA itself is more fully disclosed. Chapter 3 concludes with a brief consideration of the implications of this understanding of biological order (for which Gene-D would provide the proper gene concept) for rethinking evolutionary theory.

Chapter 4 begins with an historical analysis that parallels that of chapter 1 but this time with an emphasis on how conceptions of cancer follow from different conceptions of the basis of biological order. By the end of the nineteenth century the *Keime und Anlagen* of the neo-Kantians had become localized to within cells in general and to the ovum in particular. As repositories of the developmental potential of the whole, cells took on a certain monadic status, being both constituent parts and yet also self-contained reflections of the whole. So conceived, cancer is not determined from within, as any cell could potentially veer off in a novel direction, but rather in terms of those supracellular pathways of interaction and organization that must be the basis on which developmental destinations are realized. The monadic view of cells leads to a cellular-organizational field theory that would understand carcinogenesis to be the result of disruptions of an organizational field. A disruption might result from an environmental "irritant," leading to the misplaced expression of cellular developmental potential. However, with the phylogentic turn that takes place early in the twentieth century, a clear bifurcation in lines of thought takes place. Boveri's somatic mutation hypothesis localized the cause of cancer to *within* the cell. Cellular autonomy becomes understood not as the norm but as a kind of aberration, a malignant determination from within, as contrasted with the earlier view whereby it simply followed from the monadic nature of the cell. The fate of the monadic cell was not determined from within but rather historically and contingently according to its interactive place in successive developmental fields. With these two divergent perspectives in place, Chapter 4 reconstructs the lessons of twentieth century oncology in the form of a de facto dialectic, historically enacted. From the earliest oncogene hypothesis through the most recent work on colorectal cancer as the very paradigm of the step-wise mutation model, it will be argued that the somatic mutation hypothesis, fueled by a conflationary conception of the gene, has unexpectedly provided some of the strongest evidence on behalf of the anticonflationary, epigenesist critique.

Finally, chapter 5 will look beyond the Human Genome Project and the "century of the gene" (Keller 2001) into what appears to be the lineaments of a new rebirth of biology and its philosophy in the twenty-first century.

What Genes *Can't* Do

1

Genesis of the Gene

But however far we may proceed in analysing the genotypes into separable genes or factors, it must always be borne in mind, that the characters of the organism—their phenotypical features—are the reaction of the genotype in toto. The Mendelian units as such, taken per se are powerless.
—Wilhelm Johannsen, 1923

The full understanding of the nature of the genetic program was achieved by molecular biology only in the 1950's after the elucidation of the structure of DNA. Yet, it was already felt by the ancients that there had to have been something that ordered the raw material into the patterned system of living beings ... One of the properties of the genetic program is that it can supervise its own precise replication and that of other living systems such as organelles, cells, and whole organisms.
—Ernst Mayr, 1982

The gene is by far the most sophisticated program around.
—Bill Gates, 1994

The Gene—An Unusual Portfolio with a Compounded Legacy

The gene, to say the very least, is a most peculiar member of our current molecular menagerie. We may now speak of genes as "defined sequences of nucleic acids" with as much empirical support as when we speak of proteins, lipids, or even cells and tissues. Yet, the gene concept is routinely extended in directions that other biomedical entities are not likely to be taken. We would certainly be surprised, for example, to hear someone attribute some aspect of their personality to the fact of having their father's oligosaccharide for stubbornness. Oligosaccharides, like genes, are present in every living cell. Is it possible that two biologically ubiquitous types of molecules could be so fundamentally different that it would make perfect sense to speak of one as a determinant of, for example, one's stubborn disposition, but only humorous to ascribe as much to the other? How can it be sensible to speak of one species of biochemical but patently inappropriate and silly to speak of another as a determinant of human characteristics, let alone as the blueprint for all organisms?

The concept of the gene, unlike that of other biochemical entities, did not emerge from the *logos* of chemistry. Unlike proteins, lipids, and carbohydrates, the gene did not come on the scene as a physical entity at all but rather as a kind of placeholder in a biological theory. As the obvious etymological link with the word genesis suggests, the very sense of being a gene is that from out of which other things arise. The concept of the gene began not with an intention to put a name on some piece of matter but rather with the intention of referring to an unknown *something*, whatever that something might turn out to be, which was deemed to be responsible for the transmission of biological form between generations.

Since Watson and Crick, the gene is no longer just an abstract placeholder or a hypothetical address on a cytological structure known as a chromosome; rather, it has attained a specific physiochemical reference, i.e., as DNA. As such it is more than just a placeholder for "that which is responsible for a trait," but as an empirical entity it is also certainly other than just that which is responsible for a trait and, it will be argued, it is indeed also considerably less than that which is responsible for a trait. At once a molecule, yet also the heir to the premolecular science of transmission genetics, the gene carries a peculiarly multifaceted portfolio. Genes, like oligosaccharides, are molecular, but unlike oligosaccharides they are also conceived of as information, blueprints, books, recipes, programs, instructions, and further as active causal agents, as that which is responsible for putting the information to use as the program that runs itself.

The implications of there being one kind of physical matter, one kind of molecule—which is unique in this way, which is simultaneously physical "stuff" and information, a chemical and a program for running life—is hardly trivial. The task of explaining how simple matter can become organized into living beings, if so it does, has been one of the most fundamental and challenging questions of the entire Western philosophico-scientific tradition. Can the gene be the answer? Is it the bridge between simple physical matter and organized biotic form? It is clear from the epigraphs above that there are those who believe it is. The existence of the Human Genome Project attests to the seriousness of this belief.

Ontology Today

The attempt to explain an always messy reality on the basis of envisaging ontologically antecedent Forms or Ideas is hardly new to the western philosophical tradition, extending at least as far back as Plato. Curiously, ironically even, just as the metaphysics of transcendental forms, ideas, categories, and the like, have come to lose favor with many philosophers who increasingly set out to grapple with the unavoidably contextual aspects of truth and rightness (and for whom notions such as that of embodied and distributed cognition have become veritable watchwords), a new, putatively "naturalized" metaphysics of predeterminism has gained increasing influence—not from the lofty heights of God, Mind, Reason, or Being, but rather, as it were, from below. Such philosophers ostensibly seeking empirical moorings, as well as other investigators from the human sciences, have been increasingly looking to take their cues from biology. Biologists, in turn, having "gone molecular" and imbibed of a rhetorical soup flavored by ancillary developments in cybernetics, computer science, and linguistics, have come to adopt and adapt to a rather ethereal idiom of so-called information. And it is precisely in terms of information, with the aid of the rhetorical and metaphorical resources that this concept provides, that the gene is construed to be that which spans the chasm between physical matter and organized, biotic form. As a code, a program, a text, a blueprint, and so forth, inscribed in the one-dimensional array of DNA sequences, its meaning is understood to be self-contained. As an entity, its existence is now widely believed to be somehow temporally, ontologically, and causally antecedent to organismic becoming. The gene (or genetic program) envisaged as context-independent information for how to make an organism appears to have become the new heir to the mainstream of western metaphysics.

The Phylogenetic Turn and the History of Ontogeny

Although continuous with one long-standing tradition, the rise of the gene concept marks a radical break with another. For over 2000 years, from Aristotle through the nineteenth century, the living organism *within the confines of its own life span* had been at the center of naturalistic understanding and explanation. I will refer to a radical shift of perspective, which begins neither with Darwin nor with Mendel (although the work of both are contributing factors) but very early in the twentieth century, as the "phylogenetic turn." The intent of this phrase is to highlight the idea that as the gene and genetic program became understood to be the principal means by which adapted form is acquired, the *theater of adaptation* changed from that of individual life histories, that is, ontogenies, to that of populations over multiple generations, that is, phylogenies. As the genetic program moved to the explanatory center stage, the individual organism, with its own adaptive capacities, began to recede from view.

To adequately clarify and critically consider current usage of the word the "gene" we must locate it, as well as the associated assumptions of the phylogenetic turn in this larger context of the history of western efforts to reconcile the tension between the experiences of nature as simple physical matter and as organized life-forms. And to begin to do this we must start by exposing a shibboleth of recent philosophy of biology.[1] The shibboleth I have in mind is one that evokes the menace of creationism and insinuates that there have been only two basic organizing principles in the study of life: that of Darwinian evolution and that of creationism. Espousals of this sort continue to be ubiquitous in the philosophy of biological literature. A recent article concerned with the concepts of function and adaptation suggested that "originally, teleology was controversial because it was associated with pre-Darwinian creationist views about organisms" (Allen and Bekoff 1998). Now, in fact, the teleology of Aristotle as well as that of Immanuel Kant have both played extremely important roles in the history and advancement of our understanding of life, and dismissing them with the label "creationism" is not only misguided but also markedly misleading. It serves to create an arbitrary boundary beyond which many good neo-

Darwinians dare not cross. In another example, a recent book by a leading (and not even particularly orthodox) neo-Darwinian philosopher begins as follows: "The existence of adaptations, the fit between organisms and their environments, is one of the most striking features of the biological world. Before Darwin (1859) numerous accounts were offered to explain adaptation, the most prominent among them being the creationist account. According to this account, organisms were designed by God to fit the demands of their environments. Darwin offered an alternative proposition, the theory of evolution by natural selection."[2]

What are we to make of such a statement? The two most influential thinkers about the nature of adaptation, i.e., the fit between an organism and its surroundings, have certainly been Darwin and Aristotle. Does that mean that Aristotle was a creationist? Unless one's entire frame of reference is Victorian England and one is perhaps speaking only of certain Victorian friends of Aristotle (or really Plato), then the answer must be a resounding no! Aristotle was not a creationist; indeed, there were no references to external causation in Aristotle's biology at all. Aristotle labored to understand the nature of living beings in terms of the elements and movements from which they were constituted. He found in an organism's adapted form—that is, in its mode of existence and attunement to its environment, the organizing principle of the organism, its final cause or purpose unto itself, the for-the-sake-of-which it undergoes its formative processes. There was for Aristotle no exceptionalism, no miracles, or divine interventions. The possibilities of an adapted form were understood to be constrained by the properties of the elements of which it was composed and by an implicit principle of material conservation. It was in this interplay of the telos of the organism—i.e., that adapted way of being for-the-sake-of-which it takes on the form that it does and such material constraints—that Aristotle found the heuristic key with which to elaborate his taxonomy, anatomy, and physiology.[3]

As certain contemporaries might wish to point out, Aristotle, for whom the universe was eternal, did not have a theory for the *ultimate* origins of adapted form, that is, for the origin of species. So then in what way could he address the question of how adapted, complex, life-forms arise from nature? He did so through a theory of epigenesis. Complex, highly organized, adapted life-forms were understood to be the achievement of an ontogeny *in each and every case*. Epigenesis—the theory of the progressive, step-wise acquisition of adapted form during the developmental life history of an organism—was a hallmark and centerpiece of Aristotelian biology.

By considering what has been the relevant locus of interest for understanding how so-called simple nature can acquire complex, adapted form, one can bring into focus just what the real demarcation is between what became orthodox neo-Darwinist perspectives of the twentieth century and their most significant antecedents. The idea that the real focus ought not be upon the organism and its ontogeny but rather in processes that occur over many generations, and in relation to which individual organisms are naught but pawns, is unique to the twentieth century. The principal distinction to be made is not between creationism and evolutionism but rather between a theory of life which locates the agency for the acquisition of adapted form in ontogeny—that is, in some theory of epigenesis versus a view that expels all manner of adaptive agency from within the organism and relocates it in an external force—or as Daniel Dennett (1995) prefers to say, an *algorithm* called "natural selection." [Darwin himself (as you can see by my chronology) does not fall into this neo-Darwinian camp].[4]

Aristotle's Substantive Soul

As suggested above, Aristotle, for whom the universe was eternal, did not have a theory of the origin of species or a theory of the transmutation of species, but that is not to say that the seeds of a transmutation theory can't be located in an ontogenetically centered perspective. Aristotle's biology was a kind of functionalism. The telos of development for Aristotle was not just a matter of the reproduction of parental morphology but also that of an ability of the developing organism to adapt to shifting conditions of existence. Aristotle himself did not hold that environments were constant or that changes in the environment were "designed" for the good of the organism. As David Depew (1996) points out "this gives us new insight into why Aristotle, in acknowledging that environmental fluctuations are not always well-tuned to organisms, lays down as a matter of principle that organisms differ from inanimate objects because they are substantial beings, whose souls at the same time make them into unified forms and enable them to act appropriately to meet environmental contingencies in behaviorally plastic ways."

But what did Aristotle mean by a soul? What he didn't mean was some form of disembodied spirit or idea. What Aristotle perceived as the definitive sine qua non of being alive was physical process—that self-organized movement of heat and matter that takes in "nutriment, concocts it," and in so doing sustains itself. He referred to this as the "nutritive soul." Aristotle's nutritive soul did not tell the matter of the organism what to do. It was not a blueprint or an idea. It simply was that movement of heat and matter which, owing to its absence, distinguishes a wooden arm, albeit with all the right shapes, colors, and textures, from a bona fide living arm.

A sense of similarity between Aristotle's hylomorphic understanding of soul and much more recent descriptions of self-organizing dynamic systems is not entirely accidental. Aristotle may not have been privy to computer simulations of theoretical, nonlinear adaptive systems, but the idea that epigenesis was achieved by self-organizing movements driven by an internal orientation toward an adapted form was entirely consistent with his metaphysics. It was the nature of Aristotle's nature to inhere in purposes. Nature as a whole for Aristotle was lifelike—conceptually modeled not by the example of inertness but rather by the example of living activity.

This kind of outlook changed dramatically during the metaphysical shift that took place over the course of the seventeenth century. Nature became stripped of its capacity to self-organize as an *end unto itself*. Final cause, the for-the-sake-of-which a creature possessed the form that it comes to have, was not lost but rather relocated. Seventeenth century metaphysicians moved final cause from within nature to the mind of God. It was not by the hand of Aristotle but rather due to natural philosophers of the seventeenth century that final cause came to carry the sense of intelligent design and livings beings thereby the character of artifacts (Osler 1994). We can still see the earmarks of this legacy in the design talk of certain neo-Darwinists like Richard Dawkins (1976) and Daniel Dennett (1995) who want to tell us that it's now OK, even perspicuous, to speak in the idiom of design because we have a natural algorithm with which to do it. The most vituperative purveyors of the neo-Darwinian shibboleth are, it turns out, in closest agreement with contemporary creationists when it comes to the "as-if-by-intelligent-design" character of life. Aristotle, by contrast, and epigenesists ever since, have endeavored to explain life-forms not as artifacts designed from without but as self-organizing, "autopoietic,"[5] ends-unto-themselves.

The Antinomies of Early Modern Preformationism and Epigenesis

The advent of an explanatory crisis in biology brought forth by the new science and metaphysics of the seventeenth century was not immediate. Descartes, in particular, had no difficulty imagining that epigenesis, and even spontaneous generation, could occur simply on the basis of the new laws of matter in motion. Subsequent Cartesian mechanists, however, could no longer countenance the possibility of adapted form arising spontaneously from an unorganized nature newly construed as essentially passive. They offered, in place of epigenesis, a theory of preformation consistent with a deistic theology. In their view the embryos of all the organisms which would and could ever be had come into existence with the creation of the world and its first creatures, as so many Russian dolls, fully formed miniatures nested and encased one inside the other. Subsequent generations were deemed to "evolve" from the old on the basis of the purely mechanical unfolding and elaboration, the *inflating* really, of parts already in place.

Theories of epigenesis made a comeback during the eighteenth century, inspired by the example of Newton's discovery of gravitational force. The success of Newtonian physics meant that the natural sciences could, and did, countenance causes of action beyond the mechanics of direct collision. Where "Cartesian matter" lacked the wherewithal to become self-organized, "Newtonian matter" by contrast could yet contain some new principle, some vital force, which could account for self-organizing epigenesis (Farley 1974, Roe 1981). New epigenesists seeking to discover just such an organizing force aspired to become the Newton of natural history.

Eighteenth century attempts at addressing the problem of how nature could produce complex, adapted life-forms thus oscillated between two positions, both of which were problematic. On the one hand preformationists could not account for the production of hybrids that bore the characteristics of both parents, nor could they meet the increasing challenge of empirical observation, that is, of being able to reveal miniature adults in eggs. Epigenesists, on the other hand, couldn't solve their source-of-organization problem without referring to intangible, occult-sounding forces (*Kraft, vis essentialis, nisus formativus, and so forth*).

The Critical Solution—Teleology Turned Heuristic

What enabled nineteenth century biologists to get beyond this impasse and provide the heuristic groundwork that ultimately led to much of what became the foundation of modern biology—that is, the elaboration of developmental morphology, a histology, and embryology based on a theory of germ layers, the discovery of the mammalian ovum, the formation of cell theory and the elaboration of cellular histopathology—was not the theory of natural selection but, in effect, a renewal of Aristotelian final cause given an epistemological turn. Immanuel Kant and his Göttingen interlocutor, biologist, and ethnographer, Johann Friedrich Blumenbach, found an enabling passage through the quagmire of reconciling the fact of complex, adapted life forms with a nature construed to be mechanistic all the way down, through Kant's notion of a *reflective judgment*.

In his third critique, *The Critique of Judgment*, Kant observed that to behold a living organism unavoidably entailed regarding it as a self-sustaining, and hence internally purposeful, end unto itself. Unlike the mechanistic processes of the nonliving world which lack any internal directionality, living beings exhibit, in Kant's view, a circular causality constituting an ongoing status of being both the cause and effect of themselves. Using the example of a tree Kant observed the fact of circular causality in the following three ways:

1. As a member of species the tree is both cause and effect of other trees of its kind.

2. The tree is the cause of assimilation of nutrients into tree constituents and thus the cause of the chemical changes that these nutrients undergo and their subsequent effects as they become the matter of the tree.

3. The parts of the tree exist in a relation of reciprocal interdepend-ence—roots dependent upon shoots (foliage), shoots dependent upon roots. The parts are thus cause and effect of each other.

This kind of circular causality for Kant could not be conceived of as the result of random, mechanistic processes. We are, says Kant, com-pelled to draw reflectively on a "concept of reason," that is, one of pur-posiveness, but we do so not as a form of natural explanation but only for regulative or heuristic usage. In reflective judgment, for Kant, the subject projects "his or her own principle," the causality of reason, onto an object of nature in order to gain a conceptual handle; however, this principle, derived as it is from the subject, cannot be considered con-stitutive that is, explanatory of the object. While we must assume this teleological principle as our point of departure, that of life-forms as *Naturzwecke*, or organizational ends (purposes)-unto-themselves, we must, says Kant, nevertheless strive to account for them as far as possi-ble in the only *explanatory* mode at our disposal, that of mechanistic analysis. Whereas the proponents of the preformationism-epigenesis debate foundered in a largely fruitless effort to explain the source of this organization, Kant's principal programmatic recommendation thereafter was to take the fact of a purposeful organization as a heuristic given and proceed to explain its workings mechanistically.

So if in investigating nature we are to avoid working for nothing at all, then, in judging things whose concept as natural purpose does undoubtedly have a basis (i.e., in judging organized beings), we must always presuppose some original organization that itself uses mechanism, either to produce other organized forms or to develop the thing's own organized form into new shapes (though these shapes too always result from the purpose and conform to it). (Kant, *Critique of Judgment*).

Kant found in the Göttingen scientist Johann Friederich Blumenbach an approach to reconciling the conflict between purposeful organization and material mechanism very much akin to his own. Blumenbach had been, as were many other eighteenth century intellectuals (Lenoir 1980), highly taken with the discovery of regeneration in the fresh water polyp. The ability of the polyp to regenerate an amputated part exhibited for Blu-menbach a kind of organizational urge, or *Trieb*. But of special interest to Blumenbach was the observation that the regenerated part was always smaller than the original. What this suggested to him was that the "organizational urge" was a *material* phenomenon susceptible to physical dissipation. Blumenbach hypostatized a special inborn *Trieb*, or urge, which he specified as a *Bildungstrieb*, or formative urge. What Blumenbach understood by the *Bildungstrieb* was neither some (Platonic) soul superimposed on matter nor merely the sum result of the individual parts of an organism. It was rather a force which resulted from the *peculiar organization of a living being as a whole*. It was thus inseparable from the parts but not reducible to them. Kant said of Blumenbach:

> Yet by appealing to this principle of an original organization, a principle that is inscrutable to us, he leaves an indeterminable and yet unmistakable share to natural mechanism. The ability of the matter in an organized body to [take on] this organization he calls a formative impulse [*Bildungstrieb*] (Critique of Judgment).

And in a later writing Blumenbach states that just because we can't explain the mechanism by which the *Bildungstrieb* is brought about

> that does not hinder us in any way whatsoever, however, from attempting to investigate the effects of this force through empirical observations and to bring them under general laws (quoted in Lenoir 1982).[6]

Conception of the *Keime und Anlagen*

How does one go about empirically characterizing a *Bildungstrieb*? Can a research program be focused around an organizational force which must always already be taken as a given? Kant found the key to approaching this problem in his observations of the apparent kinship of various species suggestive of the relationship of various organisms to a common archetype and thereby the possibility of identifying some mechanism which leads from one type to another. The pursuit of a comparative anatomy, for Kant, could become a means for finding some unifying principles of nature that would reveal something about the production of organized beings. Observing gradations in the form and complexity of living taxa, Kant suggested that "an archeologist of nature" could conceive of nature producing a less complex organized being with the potential to give rise to progeny that "became better adapted to their place of origin and their relations to one another." (Kant, *Critique of*

Judgment). Kant famously referred to this speculation as a "daring adventure of reason." He even considered the possibility of a chain of common descent extending all the way from aquatic animals and marsh animals to land animals, but he dismissed this hypothesis on *empirical* grounds. Adequate evidence—intermediate forms, for example—simply was not available to support it. Still, Kant considered the transmutation of taxa within some *limited* range, based on the heritable influence of the existential conditions of organisms on their generative stock, to be highly plausible (*Critique of Judgment*).

Blumenbach, a formative influence on a whole generation of turn-of-the-century biologists at Göttingen, was the principal conduit for a Kantian-inspired research program. The strategy of seeking to elucidate the principles by means of which new life-forms are produced from some "purposefully" organized germ has been described as that of *teleomechanism* (Lenoir 1982). The idea that whole branches of the animal kingdom, such as vertebrates, were derived from some common stock of adaptive potential indeed proved to be a very powerful heuristic for triggering new research programs. At the core of the teleomechanist perspective is the idea that within the organizational form and structure of the germ, which as previously stated must be taken as a given, there is the potential, the *Keime und Anlagen*,[7] not just for a single organism but of an entire range of related basic forms, or *Baupläne*, and thus for adaptive modification. Epigenesis follows forth from the *Keime und Anlagen* of the germ, responding plastically and adaptively to the organism's conditions of existence. Epigenesis is thus construed as an adaptive process which, under conditions of sustained environmental pressure over multiple generations, may give rise to stabilized new forms, that is, to new species, which are already preadapted to the environmental pressures faced by the preceding generations.[8]

According to the teleomechanist heuristic, expressed by Kant, the possibility of these forms must already be present in the ancestral germ. The scope of the teleology here is just that of the *adaptive capacity* latent in the ("purposefully" organized) germ, an adaptive capacity which is understood to be wholly material in nature and amenable to mechanistic analysis. This brand of epigenesis as articulated by teleomechanists was at once a kind of "generic preformationism" (Lenoir, 1982). What was preformed were not particular traits but rather the possibility-space of some universe of life-forms which all share a general developmental pattern intrinsic to the *Keime und Anlagen* of the original germ. Generic preformationism imposes constraints on possible organismic form but it does so on a systematic and architectonic basis and not at the level of the individual traits of a specific organism.

Rather than foreclose or marginalize the significance of developmental adaptability, the teleomechanist's generic preformationism gave definition and importance to developmental adaptability. Indeed, it would be the selfsame principles of ontogenetic adaptation that would also be the principles of phylogenetic radiation. A clear articulation of a teleomechanistic understanding of the relationship of ontogeny to phylogeny was spelled out by Karl Ernst von Baer, a leading nineteenth-century biologist, in his influential four laws of development:

1. The more general characters of a large group of animals appear earlier in their embryos than the more special characters.

2. The less general forms develop from the most general forms, and so on, until finally the most specialized form arises.

3. Every embryo of a given animal form, instead of passing through the other forms, rather becomes separated from them.

4. Fundamentally, therefore, the embryo of a higher form never resembles any other form, but only its embryo.[9]

A depiction of von Baer's model of the relationship of ontogeny to phylogenetic radiation within a type can be seen in figure 1.1. At the center of the diagram is T, which represents the entire stock of *Keime und Anlagen* of the type. It is not meant to correspond to a particular empirical entity but rather only to a hypothetical construct—the embryological germ prior to any differentiation in the direction of a particular developmental path, i.e., the germ that still possesses the potential of the entire possibility-space of the type. Had circular representations been used instead of only branching lines and had the size of the circle been used to represent the amount of *Keime und Anlagen*, then T would be represented by the circle with the greatest diameter of any in the figure. From T two lines proceed toward C1 and C2. These lines represent two contingent developmental trajectories, two pathways in the embryology

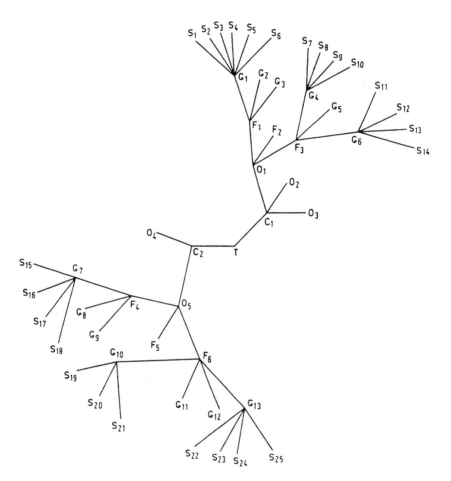

Figure 1.1
Von Baer's model of the relationship of ontogeny to phylogenetic radiation within a type; see text for details.

by means of which all the subsequent organismic forms are derived. C1 and C2 represent the first differentiations from the original stock. They represent that stage of embryological development common to all the members of Class 1 and Class 2, respectively. The circles that could best represent the stock of *Keime und Anlagen* of C1 and C2 would be smaller than that of T but larger than that of any subsequent circles in the figure. Whereas species S1 through S14 would all pass through the stage C1 common to that class and so could not be distinguished from each other at that point, they could already be distinguished from species S15 through S25, which have already taken the form of C2 common to the developmental pathway of all members of the C2 class. Neither C1 nor C2 would be construed as either necessary expressions of T nor exhaustive of the developmental trajectories latent in T. Rather C1 and C2 should be construed as historically contingent differentiations of T established under certain conditions of existence. Likewise C1 and C2 each give rise to further contingent developmental specifications, i.e., O1 through O3 and O4 and O5, respectively, which represent specialization in the developmental pathway that focuses and delimits the stock of *Keime und Anlagen* which define the possibility-space for these respective *orders*. From *order* to *family*, from *family* to *genus*, and from *genus* to *species*—each step represents a certain focusing, a certain commitment toward the specialized expression of the potential resident at the level of the previous node, with the resulting diminution of the magnitude of the resulting possibility-space latent in the developing organism. As described by von Baer's laws of development above, earlier stages of development are shared by larger and larger numbers of organisms. The earlier the stage of development, the less committed to a certain contingent direction of specialization and thus the larger the stock of the original *Keime und Anlagen* or potential possibility-space. Developing organisms, for von Baer, never pass through the adult stages of less advanced organisms, they merely pass through stages of development that are common to other members of their order, class, family, and so on. The more specialized an organism becomes the more it progressively departs from other organisms during the course of its development.

The Organizing Germ of Nineteenth-Century Biology

Von Baer's model found its most immediate expression in the elaboration of a descriptive developmental morphology (*Entwickelungsgeschichte*). By following the progressive differentiation and specialization of the forms that the *Keime und Anlagen* of a type manifest across species, embryologists obtained the means of distinguishing true homology from mere analogy. In light of von Baer's laws, the "embryological method" became the key to a rigorous methodology, albeit a descriptive one, for discerning authentic phylogenetic relationships. However, unlike the recapitulationist approach which gained vogue later in the nineteenth century under the rubric of Ernst Haeckel's "biogenetic laws," von Baer's model did not rely on any form of intuitionism and did not claim to see the lower adult organism in the embryological pathway of the higher organism. Even more tangibly, developmental morphology provided a comparative method for clarifying some of the more intricate and puzzling aspects of vertebrate, including human, embryology. The rapid success of this approach spoke well of its heuristic power. Johannes Müller's classic monograph on the embryology of the vertebrate urogenital system, for example, provides an excellent case in point.

Urogenital embryology was complicated by the appearance of apparently vestigial organs and the formation of organs from secondary association of parts which appeared first independently to serve other functions, that is, by the developmental appropriation of degenerating fetal organs. As of 1830 the accepted view was that the Wolffian duct was the common source of both the kidneys and all internal sexual organs. Müller's departure from this view was guided by a teleomechanistic perspective. Müller reconstructed the developmental path of the urogenital structures from less complex and less specialized to more complex and more specialized by examining a series of vertebrate species, including frogs, salamanders, lizards, snakes, turtles, birds, sheep, goats, cows, dogs, and humans.

He was guided by the view that the organs of all of these species arise from the same original stock of potential with the more specialized forms passing through the more simple ones during the course of their development. A closely related assumption was that if one was to claim a causal relation between an earlier and a later structure, one must do so on the basis of the observation of a sequence of structural transformations which confirm the material continuity between the earlier and later structures (Lenoir 1982).

Demonstrating the separation of the Wolffian body from the kidneys, which develop late in salamanders and at a palpable remove from the

Wolffian body, and demonstrating the absence of material continuity between the Wolffian body and the nascent kidney in chicks, mammals, and humans, Müller, employing an analogical manner of reasoning combined with structural analysis, established that the Wolffian body, or mesonephros, serves as a fetal kidney, but was *not* the source of the mature kidney. He then established that as the Wolffian body degenerates it is appropriated by the developing male genital system. However, while the Wolffian body is present in male and female vertebrates in early embryonic life, careful analysis under the watchword of material continuity, demonstrated that the Wolffian body almost completely disintegrates in female development. Where male internal organs—i.e., the epididymis, vas deferens, and seminal vesicle—are appropriated from the degenerating Wolffian body, female internal organs—i.e., the uterus, cervix, and fallopian tubes—are derived from another duct, the previously unidentified paramesonephron, henceforth known as the Müllerian duct.

Research in developmental morphology, guided by its teleomechanist heuristic, proved to be productive on a number of counts. "A central unsolved problem in this tradition, however, was that of deciding where these *Keime und Anlagen* reside and in what their nature specifically consist of" (Lenoir 1982). An important step in taking the understanding of *Keime und Anlagen* in the direction of further physical specification was achieved by von Baer in his demonstration of the mammalian ovum. In his celebrated 1827 paper, von Baer began to disclose the process of mammalian ovulation through conceiving of the ovum within its (Graffian) follicle as a kind of egg within an egg, depicting the process of ovum formation as a microcosm of development as a whole. Von Baer, in Kantian fashion, did not believe that either the ovum was prior to the follicle or that the follicle was prior to the ovum, but rather that they are the cause and effect of each other. He saw as parallel processes the formation of a smooth membrane on the surface of the follicle and the formation of a smooth membrane on the surface of the ovum. He identified the nucleus of the ovum not as a material particle but as a center of formative activity which was responsible for the formation of the granulosa layer (Lenoir 1982). While this inner core of formative activity was under the influence of the whole, the process of development itself was construed as largely centrifugal, moving outwards from center to periphery.

Modern cell theory proper began in Johannes Müller's laboratory in the 1830s with Schleiden's work on plant cells. Schwann, also a member of the Müller laboratory, quickly extended Schleiden's theory of the cellular basis of plants to animals in the course of three papers he published in 1838 (Rather 1978). Although the mood, especially among the younger generation of investigators, was shifting away from talk of *Bildungstrieb* or *Lebenskraft* and toward a more reductively materialistic language, the new cell theory was both highly influenced by, and quickly incorporated into, the teleomechanist research programs of von Baer and J. Müller.

In adapting Schleiden's theory of the plant cell to that of the animal world, Schwann had followed the pattern established by von Baer. Like the ovum and the follicle, the nucleus of the cell was perceived as a cell within a cell. The idea that morphogenesis proceeded from center to periphery introduced by von Baer became a dogma of development for Schwann (Lenoir 1982). The full potential for organismic development was deemed to reside within the cell. Cells, according to Schwann, differ along a continuum as greater or lesser realizations of their potential to form a blastoderm. If living tissue is built of cells and cells are each a repository of the entire *Keime und Anlagen* of the type, then the questions concerning both normal development and pathological manifestations become situated in a new framework. In the words of Schwann "the individual cells so operate together in a manner unknown to us to produce a harmonious whole." (Weiss 1940).

While the teleomechanist program enjoyed significant gains with the discovery of the mammalian ovum and the elaboration of cell theory, these findings would lead to a conceptual crossroads by the last two decades of the nineteenth century. In 1855, Rudolph Virchow, a former Müller student and the founding father of cytopathology, proclaimed that all cells are derived only from prior cells: *omnis cellula e cellula*. Virchow's dictum expressed the holism of the teleomechanist legacy (which could not countenance organized life arising de novo) and yet also introduced the lineaments of a new form of reductionism. Are complex organisms no more than the result of second-order interactions of autonomous cells? Is the ovum then merely a cell like any other cell, or does it also stand in a special relation to the future organism as some form of organizational forebear in a manner in which all its cellular progeny do not? This distinction was nicely expressed in retrospect by Paul Weiss (1940), whose own work sought to bring together both points of view:

We may say that the cell theory is correct: the egg is a cell and it gives rise to all the successive cell generations which contribute to the organism. But the organismic theory is likewise correct: The egg is also an organism, and it passes its organization on continuously to the germ and the body into which it gradually transforms. Only the dual concept seems to fit the facts, as we see at present. To be consistent, we should supplement Virchow's well known tenet of the cell theory: "*Omnis cellula e cellula*" by its counterpart: "*Omnis organisatio ex organisatione.*" If the former denies spontaneous generation of living matter, the latter denies spontaneous generation of organization. In admitting this, we merely paraphrase what Whitman has called the "continuity of organization." But within these specified limits the cell, even in development, is still, as Schwann has said, an individual.

In the remainder of this chapter we will see how the question of the continuity of organization, in fact, became further and further removed from that which became the mainstream of work on heredity. We will return to the question of organization in subsequent chapters.

Chopping up the *Anlagen*

The last three decades of the nineteenth century saw a proliferation of attempts to theorize about the relationships between evolution, heredity, development, and cytology. Developmental morphology after Darwin and under the influence of Ernst Haeckel, had become tinged with romantic intuitionism and subordinated to the purposes of phylogenetic reconstruction as guided by Haeckel's biogenetic law. Increasingly however, late nineteenth-century embryologists were turning toward a new experimentally oriented embryology devoted not to classification but to further revealing the mechanisms of development.

With the advance of cytological microscopy, the continuity of the nucleus and the existence of chromosomes had become empirically well established. Recalling the model of a cell-within-a-cell articulated by von Baer, it is not surprising that a pattern of centrifugal movement, this time from the nucleus outward, would be invoked. But how would the *Keime und Anlagen* of pre-Darwinian, teleomechanist thought be reconfigured to fit into a "Darwinian nucleus"? Darwin conceived of a process of gradual change requiring the presence of a range of heritable variations exposed to the forces of natural selection. Where variation in the neo-Kantian mold was construed as systematic and adaptive, Darwinian variation had to be based on the generation of many independently variable characteristics whose variants were mostly the product of random processes.

At least one way to assimilate the legacy of the teleomechanist *Keime und Anlagen* into a Darwinian framework would be to chop it into many little pieces such that natural selection could favor or disfavor small differences. Of course, in allowing for gradualist change in this manner, one must abandon the systematic and generic quality of the older model. The generic preformationism of the neo-Kantians was holistic. The adaptive potential of the *Keime und Anlagen* was necessarily understood to be a function of the purposeful organization of the whole—it could thereby not be reduced to merely a sum of its parts. When the *Keime und Anlagen* become chopped up into particles, the generic potential—i.e., the potential to produce a wide range of forms which is present in the germ as a whole—is forsaken. One returns to a kind of pre-Kantian preformationism, now in the form of a patchwork homunculus. Darwin himself conceived of heredity based upon an aggregation of particulate gemmules or pangenes which are distributed throughout the body and capable of migrating into germ cells to provide the full complement of particles necessary to produce a new organism. Darwin's model did not exclude the possibility of the inheritance of acquired characteristics in as much as such particles, produced by the active mature body, could well be influenced by life experience.

That separation between heritable factors which could or could not be influenced by life experience was put forward, as a veritable *cordon sanitaire*, by August Weismann. Weismann, an embryologist, Darwinist, anti-Larmarckian, and leading expositor of late nineteenth-century speculative biology, offered a vision aimed at a unified account of evolution and development. Weismann postulated the ironclad separation of tissue destined to be germ cells from that destined to give rise to the remainder of the body, or soma. Remaining isolated from the soma, the so-called germ plasm can't be influenced by the life-experience of individuals, thereby ruling out the possibility of passing on adaptive changes. Weismann conceived of the chromosomes as a complex hierarchy of nested determinants of form, smaller particles packed into progressively larger particles (the primary units, or "biophores," aggregated to form "determinants"; the determinants grouped into "ids"; and the ids grouped into "idants"; identified with observable chromosomes (Wilson 1893). Weismann, in effect, not only chopped the *Keime und Anlagen* into pieces but also chopped the pieces into pieces, and those pieces into pieces as well. The process of development, he believed, could be accounted for by the unequal distribution of piecemeal determinants in each successive division of the fertilized egg and its descendants. Somatic cell types, then, would be determined by the subset of heritable particulate material with which they were endowed.

Although purely speculative, Weismann's model had the virtue of offering a unified theory of development and evolution, taking its cues from the requirements of a Darwinian gradualism and a commitment to an unmitigated anti-Larmarkianism. Perhaps the popularity of these later commitments among many twentieth-century evolutionary thinkers has tended to immunize Weismann's still often celebrated reputation against what could be the more corrosive effects of empirical *dis*confirmation.[10] Weismann left an influential legacy in many ways, but his work did not lead directly to what became the predominant research program of the twentieth century.

A Bifurcation in Embryology

By the 1890s the United States was gaining importance in the world of biology. The Marine Biological Laboratory (MBL) at Woods Hole was becoming a focal point for debates among leading practitioners of the new experimental embryology. The contrasting views of two of the primary figures at Woods Hole prefigured the principal axis of opposition that came to define the subsequent course of American biology.

Charles Otis Whitman was the first director of the MBL in 1888 and founder of the *Journal of Morphology*. He rejected the autonomous-cell theory of development and considered the organization of the egg to represent the unity of the organism prior to any cell division. Rather than focusing on the primacy of the nucleus his emphasis was on the organization of the egg cytoplasm and its constitutive role in the generation of differentiated cell lineages:

The organization of the egg is carried forward to the adult as an unbroken physiological unity, or individuality, through all modifications and transformations. The remarkable inversions of embryonic material in many eggs, all of which are orderly arranged in advance of cleavage, and the interesting pressure experiments of Driesch by which a new distribution of nuclei is forced upon the egg, without any sensible modification of embryo, furnish, I believe, decisive proof of a definite organization of the egg, prior to any cell formation (Whitman 1893).

Whitman emphasized the continuity of organization from one generation to the next, with the organization of the egg cytoplasm constituting the key link between generations. This approach to development gave rise to studies which traced the origins and subsequent history of cell lineages, beginning with identifying that part of the egg cytoplasm which gives rise to particular cell lineages.

By contrast, Edmund Beecher Wilson, who was to become the leading American cellular biologist during the first decade of the twentieth century, staunchly defended the autonomous-cell-centered model of development. If all cells are equal in their developmental potential, then the cytoplasmic organization of the egg need not be granted a privileged status. Both Whitman and Wilson were keenly aware of their standpoints with respect to recasting the status of the preformationism versus epigenesis debate, and both were motivated to avoid the perceived extremes at either end (Maienshein 1987). This opposition came to be played out largely in terms of nuclear versus cytoplasmic interactions, with the former perceived as associated with the preformationist pole and the latter with that of epigenesis. Wilson thus favored a model which emphasized the influence of the nucleus on the developing cytoplasm, whereas Whitman emphasized the role of the organized cytoplasm in regulating the activity of the nucleus. Wilson's position and cognizance of its historical standpoint is ably indicated in his MBL lecture of 1893:

It is an interesting illustration of how even scientific history repeats itself that the leading issue of to-day has many points of similarity to that raised two hundred years ago between the prae-formationists and the epigenesists. Many leading biological thinkers now find themselves compelled to accept a view that has somewhat in common with the theory of prae-formation, though differing radically from its early form as held by Bonnet and other evolutionists[11] of the 18th century. No one would now maintain the archaic view that the embryo prae-exists *as such* in the ovum. Every one of its hereditary characters is, however, believed to be represented by definite structural units in the idioplasm of the germ-cell, which is therefore conceived as a kind of microcosm, not similar to, but a perfect symbol of, the macrocosm to which it gives rise (Wilson 1893).

It is strikingly clear from this quotation, seven years prior to the rediscovery of Mendel, that what Gregor Mendel's famous paper provided was not a novel concept about the nature of inheritance, but in Kuhnian terms, an examplar for use in turning this new particulate-preformationism into a research program.

Mendel's Exemplar

What Thomas Kuhn (1962) came to highlight in his 1969 postscript as the central insight of his celebrated work was the recognition of the centrality of exemplars in the history of science. Exemplars are model problem-solutions. They provide a kind of perceptual *and* methodological template for how to solve some new class of problems by learning to see that class as being similar to the one that has been solved.

The idea of particulate factors responsible for the inheritance of characteristics was not unique to Mendel; however, Mendel's paper illustrates an exemplar for how to set up an empirical practice which makes good on the concept through the ongoing production of data. In the light of an exemplar, nature shows up a certain way, its joints construed in just that manner which allows a certain kind of problem solving to ensue. In order to clarify Mendel's role in the history of the gene-concept one must look back beyond the interpretations of de Vries and Bateson to Mendel himself. Mendel, unlike his mentor C. F. Gärtner, who was principally addressing a professional audience, maintained a close relationship with the practical concerns of his Moravian countrymen. Where Gärtner and other professional breeders were occupied with questions of fertility and the changeability of species associated with the breeding of interspecific hybrids, Mendel's innovation as a breeder was to turn his attention toward *intra*specific hybrids. Mendel crossed intraspecific varieties that differed in a few discrete properties in order to clarify breeders' problems with the constancy of introgressed traits (Falk, 1995).

Gärtner, who hybridized thousands of interspecific plants, did not obtain uniform types adequate for abstracting quantifiable results. With practical motivations, Mendel established an exemplar which allowed for the appearance of discrete traits, that is, a paradigmatic approach which simultaneously defined what could count as a proper trait and disclosed its existence. He chose to work with the garden pea (*Phaseolus pisum*), in which hybrids from the mix of red and white flowering plants resulted in only red and white progeny, as opposed to working with the sweet pea (*Phaseolus lathyrus*), in which hybrids of red and white flowering plants produced red, white, *and* pink progeny (Falk 1991). Of the seven traits of the pea plant that Mendel worked with, two of these, pod color and seed coat color, were always associated with other characteristics of the plant. In constructing his exemplar, Mendel chose to classify these as "pleiotropic" effects of the same "unit-trait" and circularly to define a unit-trait as that which is transmitted as a unit in hybridization experiments (Falk 1991). What Mendel's construction of the "unit-character" provided, what his exemplar—i.e., the inheritance of unit-character traits of the garden pea—offered was a problem-solving approach by which plant breeders could make instrumentally tractable the cultivation of properly selected discrete traits. Mendel was preferring neither a universal theory of the constitution of the organism nor an account of the relationship of the genotype to the phenotype. Indeed, Mendel did not distinguish between a genotype and phenotype, nor did he have to. Rather, he adhered to a simple preformationist view of the unit-character which obviated any need for a more elaborate developmental story. As the geneticist and historian Raphael Falk has argued:

Mendel's reductionist perception is basically that of a preformationist, so that the hidden *Factoren* were for him identical with *Anlagen*—that is, preformed organs or traits. It was the preformed *Anlagen* of the traits that were transmitted in the ovules and pollen cells. On empirical as well as conceptual grounds, Mendel consciously *selected* from the traits available to him only those that "truthfully" represented the *Anlagen*, since he, like many of the early students of Mendelism thirty-five years later, did not conceive of traits as entities that are distinct from the preformed potential for the traits (Falk 1995).

Mendel's reduction of the *Anlagen* to a preformationist concept of heritable unit-characters can be understood in its practical context. His was a form of "instrumental preformationism." Where Weismann's far more elaborate model of corpuscularized *Anlagen*, being pure speculation, could not provide biologists with systematic empirical support, the appeal of Mendel's work was precisely its *exemplary* character.

The one tension that Mendel had to face in identifying the trait with the *Anlagen* for the trait was the disappearance of traits in the case of heterozygotes. He handled this, in the absence of any theory of development, by attributing to the unit-characters (*Anlagen*) the properties of dominance and recessivity. Again, the sense in which Mendel's solution can be characterized as a kind of instrumental preformationism comes to the fore. By attributing the determination of expression to the preformed *Anlagen*, i.e., to the unit-character itself, Mendel could sidestep questions of developmental interactions which were outside of his scope of interest and irrelevant to the practical applications of his exemplar. The scope and ambitions of Mendel's exemplar with its rhetorical resources became greatly expanded, however, when appropriated for use by evolutionary theorists.

While the recognized champions and proximate conduits of Mendelism, such as Hugo de Vries, William Bateson, and Wilhelm Johannsen, were themselves plant breeders and naturalists, the ultimate path by which a Mendelian gene concept became the twentieth-century successor to the nineteenth century's *Keime und Anlagen* followed the conversion of certain leading embryologists to the newly constituted science of *genetics* (Gilbert 1978). The classical gene concept emerged largely out of debates between embyologists, which culminated in the genetic studies of T. H. Morgan at Columbia University. The impact of Mendel's work on Morgan and other embryologists, however, was not direct but rather mediated by the first generation of Mendelians.

Enjoining Mendel's Discretion

At the turn of the century, Darwinism was lacking a consensual theory of heritable variation. The banner of orthodox Darwinism was held by the biometrician and eugenicist Francis Galton (also Darwin's cousin) and his student Karl Pearson. The biometrical approach to heredity and variation was largely statistical in nature and represented differences in organisms along the continuum of a bell curve. The notion of variation along a continuum, however, presented certain problems for a Darwinian theory of evolution. The classification of life-forms into species recognizes discontinuities. Heritable variations along a continuum are likely to be quickly assimilated to the common form through blending. The Dutch botanist and plant breeder Hugo De Vries, whose pangenesis model of inheritance followed the particulate tradition of Weismann, attempted to address this problem through a theory of evolutionary saltations based on large mutations. De Vries's mutations would be the cause of new species and not susceptible to assimilation by blending (Allen 1975). For de Vries the role of natural selection would be only that of eliminating unsuitable novelties and not the design of biological innovations by means of progressive incremental processes.

In England, William Bateson found that organisms distributed along an environmental gradient are prone to vary in a discontinuous manner. He too concluded that whereas the environment provided the basis for variation along a continuum, variation based directly on inheritance was discontinuous (Allen 1975). It was this vociferous debate between the advocates of continuous variation and those of discrete variation which set up and structured the rediscovery of Mendel's paper in 1900 (Carlson 1966), with the latter—i.e., de Vries, Bateson, et al.—finding in Mendel a critical resource with which to bolster their cause.

Evolutionary Mendelism and the New Particulate Preformationism

If Mendel's preformationist model can be conceived as instrumental, that is, if Mendel's work was born of "an effort to select those specific traits that segregate as distinct units, in order to achieve the empirical simplification of the problem of inheritance-as-transmission" (Falk 1995), then de Vries's intentions were of a much broader theoretical scope:

de Vries mobilized Mendel's heuristic of selecting appropriate traits that would make his experiments intelligible, and turned it into a basic conception of his theory of "intracellular pangenesis," namely, that organisms are composed of unit-characters that are distinct in transmission as much as in development (Falk 1995).

For de Vries the point of Mendelism was to vindicate a particulate-preformationist general theory of the organism which supported a macromutational[12] theory of variation and evolutionary speciation:

> According to pangenesis the total character of a plant is built up of distinct units. These so-called elements of the species, or its elementary characters, are conceived as tied to bearers of matter, a special form of material bearer corresponding to each individual character. Like chemical molecules, these elements have no transitional stages between them. . . . The lack of transitional forms between any two simple antagonistic characters in the hybrid is perhaps the best proof that such characters are well delimited units. . . . In the hybrid the two antagonistic characters lie next to each other as anlagen. In vegetative life only the dominating one is usually visible. . . . I draw the conclusion that the law of segregation of hybrids as discovered by Mendel for peas finds very general application in the plant kingdom and that it has a basic significance for the study of the units of which the species character is composed (de Vries 1900).

A central outcome of de Vries's encounter with Mendel was the modification of his notion of pangenes, which he had previously deemed to be variable in number per cell and which he then revised to pairs that segregate in the production of gametes. Mendel's studies in transmission thus served to alter de Vries's conception of the units of inheritance, but not with respect to their status as *Anlagen*. Where Mendel's reduction of the units of transmission to *Anlagen*—i.e., to the preformed units of development—was for specific practical purposes (Falk 1995), de Vries's Mendelism proffered a conception of the inheritance of particulate *Anlagen* as a universal theory. Transmission and development were fundamentally linked in de Vries's preformationism. Traits were nothing but the expansion of the inherited *Anlagen*. In order to account for why one kind of *Anlage* would be dominant and another recessive in the cases of hybrids (heterozygotes) with antagonistic characters, De Vries looked for substantive differences between them. He suggested the following: "Ordinarily the character higher in the systematic order is the dominating one, or, in cases of known ancestry, it is the older one" (de Vries 1900).

Bateson also sought to ground a discontinuous model of heritable variation in a theory of encapsulized, developmentally preformed units of hereditary transmission. His term "allelomorph" for the units of transmission captures this sense of an encapsulated piece of organismic form. He too addressed the issue of dominance in hybrids and, in order to simplify further the explanation for this phenomenon, he proposed that dominance and recessivity were really about the presence or absence of some heritable *Anlagen*. In this way, the pure-breeding recessive line simply lacked that heritable unit-character–allelomorph which shows up as dominant in hybrids (heterozygotes) whose other parent comes from a pure breeding line that possesses it. What this does not explain is why a trait would fail to appear or appear in a highly attenuated form when the allelomorph is present. To explain (away) such deviations from Mendelian expectations without recourse to complicating the story by reference to developmental interactions, the environment, and so forth, the terms *penetrance* and *expressivity*, were introduced, which simply turned phenotypic variability into *intrinsic propensities* of the allelomorphs.[13]

Johannsen's Critique of the New Preformatonism—Origins of the Phenotype-Genotype Distinction

The terms "gene," "genotype" and "phenotype" were introduced by the Danish botanist Wilhelm Johannsen, but his contribution to the advancement of the gene concept ran far deeper than terminology. *Keime* and *Anlagen* in nineteenth-century biology were concepts that pertained first and foremost to development. Holistic and teleological when taken together, they were not readily amenable to a reductive analysis. The teleomechanist program refined its understanding of inheritance to the extent that it established the continuity of the cell. The Kantian heuristic required that the wherewithal for producing new forms be always already contained within the potential of the germ. An explanation for the appearance of new species (past or future) would thus be a further extension of its theory of development, i.e., a new and more adaptively specialized expression of the potential of the germ.

The question of transmission across generations only became an important topic unto itself when the view arose that evolutionary changes were based at least in part upon novel and possibly fortuitous variations in the germ. Darwinian natural selection requires that progeny of a species vary in what they contain in their egg, that is, in what is transmitted from their parents. In attempting to formulate a way to conceive of the contents of the egg such as to be consistent with a gradualist theory of evolution based on natural selection, Weismannians and biometricians alike required a conception of the determinants of organismic form which would allow for all kinds of minor variations of character that can be packaged within the egg. Johannsen (1911) considered this legacy to be one steeped in the popular experience of heredity as the inheritance of personal properties, as, for example, in the inheritance of an estate. In later writings (1923) he extended his critique to what he called the whole "morphological tradition" in which he included, not just Darwin, Weismann, and Galton, but also Mendel, Bateson, and de Vries. What he meant by the morphological tradition was the penchant for believing that units of inheritance were chunks of morphology, as patently expressed in Bateson's term "allelomorph." However, the morphology of an organism is the contingent result of many factors interacting over time. Johannsen coined the terms genotype and phenotype and distinguished between them in order to depart from this morphological (and preformationist) legacy and establish, he felt, the grounds for a proper science of the inheritance of the genotype. It is precisely the conflation of the phenotype—a product of environmental-developmental interactions—with the inheritance of Mendelian units which constituted a new brand of preformationism.

What Johannsen called for in distinguishing between the genotype and the phenotype was a separation of the inheritance of Mendelian units from development, thereby constituting the study of genetics as an independent discipline. All of the theories of inheritance from Darwin through Bateson and de Vries had confused the two. Genetics emerges with Johannsen as a science of the acquisition of the genotype and *not* of the phenotype. Heredity was not about the passing on of *properties*, which were always historically acquired and developmentally contingent, but rather about the presence of identical genes in ancestor and descendent. Although Johannsen had no specific suggestion for the (physical) nature of genes, he recommended treating them as chemical-like in their *ahistorical* nature:

The genotype conception is thus an "ahistoric" view of the reactions of living beings—of course only as far as true heredity is concerned. This view is an analog to the chemical view, as already pointed out; chemical compounds have no compromising ante-act, H_2O is always H_2O, and reacts always in the same manner, whatsoever may be the "history" of its formation or the earlier state of its elements. I suggest that it is useful to emphasize this "radical" ahistoric genotype-conception of heredity in its strict antagonism to the transmission-or phenotype-view (Johannsen 1911).

Having distinguished between development and inheritance Johanssen had no need to interpret the gene in a reductivist-preformationist fashion. He recommended that the genotype be understood along the lines of the German term *Reaktionsnormen* used by Woltereck to refer to the full range of an organism's potential (1911). The phenotype of an organism for Johannsen is the product of the *whole* genotype reacting to the environmental conditions of its development. Phenotypes can be seen to vary along a continuum because the *Reakionnormen* of the genotype are capable of plastically adapting to variant conditions. Genotypes vary discretely, but the consequences on phenotype are realized at the level of the *Reaktionnormen* as a whole:

Hence the talk of the "genes for any particular character" ought to be omitted, even in cases where no danger of confusion seems to exist. So, as to the classical cases of peas, it is not correct to speak of the gene—or genes—for "yellow" in the cotyledons or for their "wrinkles,"—yellow color and wrinkled shape being only reactions of factors that may have many other effects in the pea-plants (Johannsen 1911).

From Cytoplasmic *Anlagen* to Morgan's Conversion

Johannsen clarified the conceptual basis for an independent science of genetics, but it was T. H. Morgan who turned it into an actual research program. Morgan, unlike the founders of modern Mendelism, was not a plant breeder or evolutionary naturalist but rather an embryologist. He was steeped in the controversies introduced in previous sections ("Chopping up the *Anlagen*" and "A Bifurcation in Embryology") pertaining to how the potential of the organism is distributed between nucleus and cytoplasm and between egg cell and progeny cells. The fact that the course of this history does not lead in a logically compelled or conclusive way to the gene-centered paradigm is what makes it interesting and important. Embryology did not culminate in genetics—rather, Morgan converted to the practice of the new discipline, leaving the unresolved problematics of embryology to fare for themselves (Allen 1985, Darden 1991). As embryologists fractured the cell into nucleus and cytoplasm, so the life sciences fractured into a center and periphery, with genetics becoming the center and with the legacy of developmentally (and organizationally) oriented biology relegated to the periphery. Philosophically, it will be important to see how many central problems were banished to the margins and yet naively thought to be solved (or almost solved) in the name of the gene. Johannsen, with much perspicuity, anticipated the likely misconstrual of the genetic perspective. After reconstructing this tortuous journey I will return to take a closer look at Johannsen's insightful perspective.

The earliest experimental attempt to locate the *Anlagen* at a subcellular level began not with the chromosome or even nucleus but rather with the nineteenth-century hypothesis of "cytoplasmic anlagen" put forward by Wilhelm His (Gilbert 1978). C. O. Whitman, following His, believed organismal development could be analyzed by tracing the path of cell lineages from their origins in cytoplasmic *Anlagen*. Not long after His, Nägeli proposed the existence of an "idioplasm," later elaborated by Weismann, consisting of nested, hierarchically organized, particulate units that directed development. Nägeli was nonconmittal, however, as to where the idioplasm would be located. With the accumulated cytological work on chromosomes and the strong impression made by the observation that chromosome number and morphology remain constant across generations, Oscar Hertwig, Weismann, Kölliker, and others postulated that the chromosomes were the site of the putative idioplasm (Sapp 1987).

That party lines among embryologists in the 1890s were partitioned along the boundary of the cell nucleus was largely contingent. Surely some form of nucleocytoplasmic egalitarianism could also have found a place within the logical space of possibilities. However, the nucleus versus cytoplasm divide came to define a rather intractable opposition over what was going to count as the proper stuff of heredity.

Wilhelm Roux, following Weismann, championed a "mosaic hypothesis," which, had it been successful, would have enabled the nuclear-idioplasm theory to unite development and inheritance within a single, particulate-preformationistic model. According to the mosaic hypothesis, cellular differentiation and organismal development are regulated by the uneven distribution of idioplasm at each stage of cell division. In this view only germ cells contain the full complement of *Anlagen*. The *Anlagen* then becomes differentially partitioned into daughter cells during somatic cell division. Ultimately each cell only receives those *Anlagen* required for its terminal state of differentiation. Roux had claimed experimental confirmation of his theory when, using a hot sterilized needle, he ablated one of the blastomeres of the two-cell-stage frog embryo. The remaining blastomere continued to develop, albeit abnormally, resulting in something like half an embryo. According to the mosaic theory, the remaining blastomere would contain only that idioplasm capable of producing half an organism (Allen 1975).

Hans Driesch, working in Naples in the 1880s and 1890s, attempted to reproduce these findings, using an agitational method for separating the two blastomeres of the sea urchin. Contrary to Roux's findings, the remaining blastomere was fully capable of producing a normal, albeit smaller, sea urchin larva (Allen 1975). Driesch countered Roux's mosaic model with his theory of the developing organism as a "harmonious equi-potential system." According to this theory, cells retain the full potential of the organism, gaining their developmental specificity not through directives from within but rather externally and relationally with respect to their position in the developing organism and the influence of environment. Driesch, in effect, proposed a nascent developmental field theory. Although Roux was disinclined to concede the point, it became generally accepted that his results were due to the residual effects of the ablated cell matter and that Driesch's results were truly indicative of the absence of mosaic nuclear division.

By 1910 Morgan could state unhesitatingly:

We have every evidence that in embryonic development the responsive action of the cytoplasm is the real seat of the changes going on at this time, while the chromosomes remain apparently constant throughout the process (Morgan 1910).

Had the mosaic model been successful, as suggested above, many problems would have been avoided. The idioplasm, presumably located in the nucleus, would have been seen to "represent" the organism as a whole and to provide the mechanism for controlling development. The process of heredity and the process of development would have been substantially united. Morgan was never sympathetic to this model because it smacked too much of old-fashioned preformationism.

If mosaic division does not explain development then one can question whether the idioplasm in fact represents the unity of the organism at all. For those who held to an autonomous cell theory, notably E. B. Wilson, it did. Wilson felt that every cell contained the full representation of the organism in its nucleus and that the nucleus directed the developmental epigenesis which was played out in the cytoplasm. How the nucleus could do this while remaining ostensibly constant throughout the organism was a question he could not answer. Driesch and Whitman by contrast believed that only the whole organism represented the unity of the whole. Individual cells did not constitute autonomous units but rather were centers of action determined by their position in the whole (Sapp 1987). From the latter point of view heredity, in any full sense, could not be explained solely on the basis of the nucleus. Subsequent debate among embryologists was concerned with the relative importance of the cytoplasm versus nucleus in heredity but also and ultimately with what would count as heredity.

In 1903 Sutton identified the processes of Mendelian segregation (of organismal traits) with those of chromosomal reduction (in cytological observation), thus again nominating the chromosomes as the putative site of Mendelian factors. E. B. Wilson and Theodor Boveri endorsed Sutton's position within a year (Morgan 1910) and proceeded to perform new experiments to show the primacy of the chromosomes in inheritance. Wilson, working with protozoa, offered evidence in support of the role of the chromosomes in regeneration.

Boveri provided key evidence in favor of the identification of chromosomes with Mendelian factors (the Sutton-Boveri hypothesis) by demonstrating the "individuality" of chromosomes, i.e., that chromosomes were not homogeneous in their biological effects. By producing dispermic sea urchin eggs, ova fertilized with two sperm, he could acquire blastomeres with different complements of chromosomes. Separating these at the four-cell stage Boveri could show that they possessed different developmental potentials ostensibly based on their differential possession of chromosomes (Gilbert 1978). Further, Boveri attempted to address the cytoplasm versus nucleus debate directly by use of "merogony" experiments. Working with enucleated echinoderm eggs, Boveri would replace the maternal nucleus with that from sperm of a phenotypically distinguishable variety. The relative impact of the cytoplasm versus that of the transplanted nucleus on subsequent development could then be observed. Boveri reported that it was the phenotypic effects of the sperm which prevailed (Gilbert 1978, Sapp 1987). Adapting methods reflective of his long interest in the biology of regeneration (Allen 1985), Morgan examined the effects upon subsequent development after removing cytoplasm from an unsegmented ctenophore egg. He reported that loss of cytoplasm despite an intact nucleus resulted in a deformed embryo. Morgan was skeptical of the findings of Boveri's merogony experiments and during the course of extensive efforts to reproduce them found increasing evidence for the primacy of the cytoplasm, *not* the nucleus, in early development.

Whether Mendelian factors reside on the chromosomes was not the only pressing question for leading embryologists during the first three decades of the twentieth century. In the absence of something like a mosaic hypothesis which united idioplasm with development and with ample evidence supporting the developmental efficacy of the cytoplasm, one also had to address questions pertaining to the scope and explanatory significance of identifying Mendelian factors with chromosomes. If the Sutton-Boveri hypothesis were vindicated (as of course it was), would that mean that chromosomes contain a full representation of the organism as Weismann and Wilson would have it? Would chromosomes thus be both necessary and sufficient for organizing the full development of the organism? This conclusion does not necessarily follow. Alternatively, the chromosomes–Mendelian factors may contain only certain, and maybe even only superficial, features of the organism, depending on the larger context provided by the cytoplasm, the totality of the organism, and the environment for successful deployment.

Jacques Loeb, a leading expositor of mechanistic materialism in biology, suggested in 1916 that "the unity of the organism is due to the fact that the egg (or rather its cytoplasm) is the future embryo upon which the Mendelian factors in the chromosomes can impress only individual characteristics, probably by giving rise to special hormones or enzymes" (Sapp 1987). Edwin Conklin, who followed Whitman and

Lillie in cell lineage studies and was a lifelong advocate of the importance of the cytoplasm of the egg, argued in 1908 that "the characteristics of the phylum are present in the cytoplasm of the egg cell" (Sapp 1987). Even Boveri in 1903 had attributed to the cytoplasm the generic features of the organism that provided the developmental context in which the individuating effects of the nucleus could and would become expressed (Sapp 1987). Morgan, in 1910, rephrased the traditional opposition between preformationism and epigenesis in terms of the "particulate theory of development" versus the theory of "physico-chemical reaction." He suggested of these that:

The particulate theory may appear more tangible, definite and concrete because it seems to make a more direct appeal to a material basis of development and heredity. The theory of physico-chemical reaction may seem more vague and elusive, since the responses and reactions to which it must appeal are as yet little known. But this distinction is not one of much importance. For the particulate theory requires as elaborate a series of processes or changes to account for the distribution of the postulated particles and their development into characters as does the reaction theory itself, and on the other hand the reaction theory may rest its claims on as definite a physical or material basis as does the other view. One theory lays emphasis on the material particles of development, the other on the changes or activities in the same material . . . Whichever view we adopt will depend first upon which conception seems more likely to open up further lines of profitable investigation, and second which conception seems better in accord with the body of evidence at hand concerning the processes of development . . . it may be said in general that the particulate theory is the more picturesque or artistic conception of the developmental process. As a theory it has in the past dealt largely in symbolism and is inclined to make hard and fast distinctions. It seems to better satisfy a class of type of mind that asks for a finalistic solution, even though the solution be purely formal. But the very intellectual security that follows in the train of such theories seems to me less stimulating for further research than does the restlessness of spirit that is associated with the alternative conception (Morgan 1910).

Morgan's sympathies had been with the physicochemical, or epigenetic, side of the coin. He contributed over many years to the evidence supporting the importance of the cytoplasm in the developmental acquisition of heritable characteristics. It was, after all, the cytoplasm, not the chromosomes, that appear to undergo differentiation over developmental time. Morgan's movement to the particulate camp, while reflective of his recognition that sex is chromosomally and not, as he had previously suspected, environmentally determined in *Drosophila* (Gilbert 1978), by

no means presupposed a rejection of all the evidence for the role of cytoplasm and physiochemical epigenesis in development. The success of
Morgan's "Fly Room" research mapping *Drosophilia* chromosomes
marked the partitioning of embryology into what would count as work
in heredity and take center stage and what would not count as heredity
and, hence, be marginalized.

At the opposite conceptual pole from that of a particulate theory
of hereditary factors lies the continuing tradition of holistic theories
attempting to explain orderly patterns of reaction in terms of biological
organization and/or developmental fields. An early attempt to articulate
a field theory was made in 1924 by C. M. Child, who sought to find biological organization in gradients of metabolic intensities. While not considered adequate by most contemporaries, the field concept was adopted
by such investigtors as Julian Huxley, Gavin de Beer, and Paul Weiss in
years to come (Sapp 1987). When Morgan turned to the chromosome,
he did not attempt to subsume the questions of development and organization under a particulate model. Rather, he followed the guidance of
Johannsen in rejecting the preformationist legacy of interpreting
Mendelian factors as *Anlagen*. By bracketing the questions of development and thus the role of the cytoplasmic-organizational context in
realizing the acquisition of hereditary characteristics, Morgan could, following his first desideratum stated above of "opening up profitable lines
of investigation," establish an independent science of the genotype.
Morgan did not claim that a science of the genotype was tantamount to
a science of the phenotype, and yet, in 1926 he stated: "Except for the
rare cases of plastid inheritance all known characters can be sufficiently
accounted for by the presence of genes in the chromosomes. In a word
the cytoplasm may be ignored genetically" (Morgan 1910). Unless newly
tempted by particulate preformationism, Morgan's intent was to establish the acquisition of genes and genotypes as the *definition* of what
counts as heredity, that is, to separate by definitional fiat the inheritance
of genes from the developmental context and mechanism which allow
heritable traits to appear. As Sapp (1987) has detailed, the fixation of
the genetic meaning of heredity was hardly uncontroversial or apolitical.
Conklin in 1919 protested: "Development is indeed a vastly greater and
more complicated problem than heredity, if by the latter is meant merely

the transmission of germinal units from one generation to the next" (Sapp 1987). Speaking perhaps on behalf of the accumulated sentiments of that sector of the life sciences whose concerns had become partitioned *outside* of the realm of inheritance, Ross Harrison in 1937 suggested:

> The prestige of success enjoyed by the gene theory might easily become a hindrance to the understanding of development by directing our attention solely to the genome, whereas cell movements, differentiation and in fact all developmental processes are actually effected by the cytoplasm. Already we have theories that refer the process of development to genic action and regard the whole performance as no more than the realization of the potencies of the genes. Such theories are altogether too onesided (Sapp 1987, p. 50).

Morgan's willingness to conceptualize his work on the genetics of *Drosophila* in terms which marginalized developmental concerns was bolstered by several factors. Morgan's Fly Room laboratory at Columbia received enthusiastic support from zoology department chair E. B. Wilson, whose advocacy of the cellular autonomy-nuclear idioplasm perspective Morgan once took issue with. Morgan benefited financially from rising industrial interest in agricultural genetics.

Ideologically, Morgan, owing in some measure to the influence of Jacques Loeb, had adopted a staunchly mechanistic materialist outlook and found in genetics a considerably more tangible model of biological causation than that found in the lexicon of epigenetic field theories (Allen 1985). The constitution of a science of heredity which presupposes the bracketing out of mechanisms involved in the realization of organismal characteristics is, as Falk (1995) has argued, a kind of instrumental reductionism. The fruit of Morgan's instrumental reductionism, in its advancing a theory of the chromosomes and in providing a handle for grappling with certain kinds of inheritance in animals and plants, can't be denied. But it is also the case that the propensity for the instrumental reductionism of Morgan's genetics to spawn naively reductionist-preformationist progeny has also had its consequences, both at the level of pernicious social ideologies and, I will argue, conceptual confusions.

Genetics arose as a discipline proper when Morgan found that in *Drosophilia* factors affecting eye color, body color, wing shape, and sex segregated together with the X chromosome (Gilbert 1978). At that point he turned to Johannsen for the conceptual framework with which to understand this. Following Johannsen, Morgan chose to separate traits from genes as well as phenotypes from genotypes and to use traits as *markers* for the hereditary entities. As Falk (1995) points out:

Not all saw the instrumental, tactical aspect of Morgan's mechanistic materialism. Many future developments of genetics, and especially offshoots of it, may be traced back to the conceptual rather than heuristic interpretation of the gene as determinants of characters. The term "genes for" . . . became, notwithstanding Johannsen's reservations, a decisive factor in our genetic thinking.

Did Johannsen Get It Right?

Whatever the full configuration of motivations were for Morgan, it is evident that once he turned toward the path of genetic analysis he did not look back. The same cannot be said for Johannsen, whose formulation of the genotype-phenotype distinction enabled Morgan to constitute a practice of genetics distinct from development. In his 1923 remarks about "units in heredity," Johannsen raised three lines of questioning which merit a place in this ongoing conversation.

Johannsen's first question pertains to the extent to which the fundamental features of the organism are segregable as Mendelian units.

Certainly by far the most comprehensive and most decisive part of the whole genotype does not seem to be able to segregate in units; and as yet we are mostly operating with "characters," which are rather superficial in comparison with the fundamental Specific or Generic nature of the organism. This holds good even in those frequent cases where the characters in question may have the greatest importance for the welfare or economic value of the individuals.

We are very far from the ideal of enthusiastic Mendelians, viz. the possibility of dissolving genotypes into relatively small units, be they called genes, allelomorphs, factors, or something else. Personally I believe in a great central "something" as yet not divisible into separate factors. The pomace-flies in Morgan's splendid experiments continue to be pomace flies even if they lose all "good" genes necessary for a normal fly-life, or if they be possessed with all the "bad" genes, detrimental to the welfare of this little friend of the geneticist (Johannsen 1923).

Johannsen is posing a question as to the scope of genetic decomposability. Segregation is the evidence for decomposability in inheritance. But does the fact of the segragatability of some characteristics, require that all inheritance is decomposable? Clearly there is no *logical* necessity that one follows the other. It could be that, as Johannsen intimates, only certain and perhaps comparatively superficial aspects of the organism are decomposable in their manner of inheritance. Nor would this be inconsistent with the practice (and success) of Mendelian genetics as a form of instrumental reductionism in areas such as agriculture and medicine. Breeders continue to benefit from the ability to reinforce selectively certain desirable features of crops without concern for the decomposibility of those central features of plants that most or all plants have in common. And, just as in the case of Morgan's flies—which, despite however many genetic aberrations they enjoy, continue to be flies—physicians can (and do) benefit from the ability to identify heritable diseases and genetic syndromes by use of traditional Mendelian pedigrees (and penetrance and expressivity "fudge" factors). And this is so even if the central features of being human (let alone mammal) never do enter into the realm of Mendelian segregation.

Although a critique of the limits of the scope of decomposability does not undermine the intentions of genetics as an instrumental reductionism, the same does *not* hold for the intentions of a constitutive reductionism. By this latter statement I simply mean a theory which treats the organism as *fundamentally* decomposable. If the word "gene" is meant to denote an entity that is causally responsible for a piece of the phenotype and if genetics taken in the vein of constitutive reductionism requires that the whole phenotype is explicable in terms of genes, then there cannot be a limit on the scope of genetic decompossibility for fear that the whole enterprise might hit a rocky shore. One might be tempted to suggest that with the benefits of molecular hindsight we can disregard Johannsen's concerns because we now understand genes to be segments of DNA whose decomposability is inscribed in the start-and-stop codes that demarcate genetic reading sequences. But this would be tantamount to a kind of category mistake.

Neither the basis for an instrumentally reductionist genetics nor a constitutively reductionist genetics follows from the structure of DNA itself, which does not bring with it either an epigenesist or a preformationist name tag. Nor does the ability to correlate certain instances of classical segregation of traits with certain DNA sequences imply that the remainder of DNA is going to be amenable to any form of classical reductionist analysis with respect to its relationship to a phenotype. When DNA segregates during meiosis, it is not the phenotype that is segregating; thus, the structure and dynamics of DNA do not address the question of the decomposability of the phenotype. The very same instrumental critique of classical genes can be replayed with DNA in mind. On the basis of observed patterns of Mendelian inheritance, we can treat certain molecular genes as if they preformationistically determined phenotypes [something like a BRCA1 or cystic fibrosis gene would be examples (see below)] while understanding that at the mechanistic-causal level of explanation DNA participates in the construction of the phenotype in a manner not amenable to reductionist decomposition. We have no reason to rule out the possibility that the more "species-typic" and "genus-typic" characteristics which are clearly heritable but never seen to segregate are based on structures hierarchically above (or simply other than) genes—e.g., chromosomal organization, membrane organization, metabolic dynamics, and so forth.

Johannsen's second question pertains to exactly those genes that do behave in a classically Mendelian fashion. With respect to these he wonders what the relationship is between so-called dominant and so-called recessive alternatives, or, for that matter, between alternative genes at all, endorsing our current formulation of "multiple allelos" as "different states in the same locus of a chromosome."

When we regard Mendelian "pairs," Aa, Bb and so on, it is in most cases a *normal* reaction (character) that is the "allele" to an *abnormal*. Yellow in ripe peas[e] is normal, the green is an expression for imperfect ripeness as can easily be proven experimentally, e.g., by etherization . . . The rich material from the American *Drosophilia*-researches of Morgan's school has supplied many cases of multiple allelisms—most of all of them being different "abnormalities" compared with the characters of the normal wild fly . . . To my mind the main question in regard to these units is this: Are experimentally demonstrated units anything more than expressions for local deviations from the original ("normal") constitutional state in the chromosomes?

Is the whole of Mendelism perhaps nothing but an establishment of very many chromosomal irregularities, disturbances or diseases of enormously practical and theoretical importance but without deeper value for an understanding of the "normal" constitution of natural biotypes (Johannsen 1923)?

It is more than noteworthy that although Johannsen was a critical link in the chain that led to classical genetics, students of genetics to this day understand Mendel's exemplary work on the pea to mean that a genetic locus typically houses several qualitatively different traits as opposed to various deviations from a single standard capacity. Similarly, what percentage of that public for whom "a gene for blue eyes" is famous and canonical understand that said gene possesses no substantive capacity for producing blue color but only an *in*ability to produce that brown pigment which may mask the blue which is already there?

While fantastical notions of alleles as alternative qualitative traits bespeak a caricaturesque preformationism, the idea of alleles as deviations from a norm is quite compatible with an instrumentalist reductionism. Deviations from a norm may become advantageous when the context that determines what is normal becomes shifted. For example, consider agricultural applications where the locus of normality shifts from that of ecological fitness to that of commercial value. Navel oranges and other fruits are bred for seedlessness and become unable to reproduce sexually. High-yield grains have lost metabolic versatility. Breeders can take advantage of abnormal allelic variants by imposing new conditions of normality. In medicine the context of normality is generally anchored in the taken-for-granted presuppositions of a culture. The aura of objectivity becomes problemized with, for example, the emergence of groups such as the hearing impaired, who constitute themselves as new focal points of normativity, that is, as alternative forms of life with distinctive beliefs about what should count as normal. New alleles that bring about qualitative differences in the phenotype generally do so because they entail the loss of some biochemical activity, which results in a regrouping at a higher level of organization, be it organismal or organismal *and* sociocultural.

Finally, Johannsen broached the question of the cytoplasm. His remarks on this topic are brief but not without interest:

Chromosomes are doubtless vehicles for "Mendelian inheritance" but *Cytoplasm* has its importance too . . . Gametogenesis with chromosome-reductions, accompanied by reformation and, as it were, partial rejuvenescence of cell-structure, must in some way act as if especially organized for *obliterating* the individual's personally "acquired characters," which as a rule totally disappear in sexual reproduction . . . Cytoplasm is perhaps more prone to "memory," Jollos's experiments with Infusoria for instance seem to suggest such a case (Johannsen 1923).

Johannsen reproduces in these remarks the central distinction he wishes to make between the genotype and phenotype. The genotype, taken as a whole, confers an ahistorical potential for a full range of phenotypes where the phenotype reflects the genotype in the context of the ongoing result of cumulative experience. The chromosomes, which clearly stand in a special relationship to the genotype, undergo a kind of "rejuvenescence" during gametogenesis, which serves to wipe the slate clean of historical experience. The cytoplasm, by contrast, appears to be capable of responding to the conditions of lived existence and of retaining the lessons of experience as a kind of memory. In the case of the Jollos experiments that Johannsen refers to, it was found that protozoa exposed to extreme conditions may undergo physiological adaptations and retain such adaptations for many generations in the absence of those conditions. Ultimately the protozoa were found to be capable of reverting back to the nonadapted state. In as much as the protozoa appeared to adapt and revert on a population-wide basis and not on the basis of the clonal selection and expansion of a mutant cell, the phenomenon displayed the character of an epigenetic cellular memory. Johannsen's ascription of memory to the cytoplasm, prompted it appears by Jollos's work, recalls both Morgan's earlier emphasis on the role of the cytoplasm in ontogenetic differentiation and development, as well as that of Driesch, Boveri, Whitman, and Conklin on the role of the cytoplasm in setting up the more generic properties of the organism. It should also be noted that the question of the rejuvenescence of the chromosome derived from differentiated cells is exactly the central technical challenge of human animal cloning, and especially in light of recent revelations of the widespread difficulties in producing healthy animals through such means,[14] it is a question in relation to which the jury is still out.

Johannsen's demarcation of the genotype from the phenotype provided the conceptual groundwork for Morgan to use phenotypic features as markers for underlying genotypic realities. In so doing he could evade the pitfalls of the morphological tradition, i.e., of the reductionistic preformationism with which Morgan would have no truck. In the practice of instrumental reductionism, genes are not construed as particles of the phenotype; rather, aspects of the phenotype are used as markers of genes *as if* they were directly determined by genes in order to provide the window needed to develop a science of the genotype. Classical genetics enjoyed its formative stage in Morgan's Fly Room where genes materialized as alleles at chromosomal loci which could be mapped with respect to their chromosomal address and linkage neighborhood. However, in the wake of the fruits of the instrumentalist program, the clarity obtained in Johannsen's reflections—the conceptual high-water mark of the classical gene concept—quickly became muddied.

Johannsen's model made possible, not only a productive application of instrumental reductionism, but also, and inseparably, a lens with which to resolve its meaning.

The necessary complement to the instrumental preformationism afforded by Johannsen is, I will argue, an epigenesist research program with which to reveal its biological meaning. Following Johannsen's vision, the genotype *as a whole* confers the potential for a wide range of phenotypes with an ability to adapt to the needs of the particular circumstances of existence. The immediate context that determines the way in which the potential of the genotype is realized is the organizational structure of the cell-organism—which we can now envisage at the level of chromosomal, membrane, cellular, supercellular organization, and metabolic dynamics—indeed, all that lies beyond the one-dimensional array of coding nucleic acid sequences. And the cytoplasm of the organism, as inferred by Driesch and his holistic successors, is immediately responsive to the larger environment. Now, the chromosomes, as Johannsen anticipated, may well undergo a form of rejuvenescence, but the cytoplasm of the egg is, as Whitman, Conklin, and Lillie held, a very likely candidate for retaining historical (generic? species?) memory. Johannsen's instrumental reductionism and genotype concept require that the genotype at birth is conceived independently of any cytoplasmic historical memory. But given the holistic nature and pluralistic potential of Johannsen's genotype, the achievement of the phenotype must be the result of an epigenesis within which chromosomal, cytoplasmic, and environmental constituents become mutually and reciprocally causal, instructive, and determinative of the outcome. Ironically, as the means for elucidating the ahistorical chemical features of the genotype emerged, that embryological tradition that was best equipped to provide the necessary complement for elucidating the context-specific interactions which actually produce a phenotype became increasingly marginalized.

Information by Conflation

The insights that allowed genetics to emerge as an independent discipline included insight into its own limitations. But the victory of genetics in securing the mantle of heredity for its sole possession (Sapp 1987) left little room for humility, conceptual or otherwise. The self-understanding of Johannsen's genetics, i.e., as that of an instrumental reductionism, gave way to a less-reflective disciplinary juggernaut. If geneticists were not going to pursue the biology of the phenotype by way of a theory of epigenesis, the alternative, other than a return to old-fashioned morphological preformationism, would have to be along the lines of a new preformationism that locates within the gene its own instructions for use. The idiom, if not the substance, for describing this was soon found in the jargon of "codes and information" which began to surface in the 1940s but hit pay dirt after the Watson and Crick breakthrough in 1953. Molecular genetics emerged as essentially that science that would explain, in physiochemical terms, how the genotype contains within itself the instructions for making an organism. Its recruits arrived largely from the shores of physics and chemistry and included among its ranks many for whom even a current knowledge of the cell was more biology than deemed necessary for the putatively information–encryption-theoretic task at hand.

The rhetoric of the gene as code and information, so familiar now as to resemble common sense, turns on, I will argue, a conflation of two distinctly different meanings of the gene. When scientists and clinicians speak of genes for breast cancer, genes for cystic fibrosis, or genes for blue eyes, they are referring to a sense of the gene defined by its relationship to a phenotype (i.e., the characteristics of the person or organism) and not to a molecular sequence. The condition for having a gene for blue eyes or a gene for cystic fibrosis does not entail having a specific nucleic acid (DNA) sequence but rather an ability to predict, within certain contextual limits, the likelihood of some phenotypic trait. What molecular studies have revealed is that these phenotypic differences are not due to the *presence* of two qualitatively different capabilities, but rather the *absence* of the ability to make the so-called normal protein. Accordingly, there is no specific structure for the gene for white flowers or the gene for blue eyes or the gene for many diseases because there are many structural ways to be *lacking* the usual resource. The white flower, the blue eye, the albino skin, the cystic fibrosis lung are all the highly complex results of what an organism will do in the absence of certain normal molecular structures.

It continues to be useful, in some contexts, to employ this usage of the word "gene." To speak of a gene for a phenotype is to speak as if, but *only* as if, it directly determines the phenotype. It is a form of preformationism but one deployed for the sake of instrumental utility. I call this sense of the gene—Gene-P, with the P for preformationist (see Figure 1.2). Genes for phenotypes, i.e., Genes-P, can be found, generally—and as Johanssen surmised—where some deviation from a normal sequence results with some predictability in a phenotypic difference.[15] In the absence of the normal sequence necessary for making brown eye pigment, blue eye color results. Any absence of this brown eye–making resource will thus count as a gene for blue eyes. Blue eyes are not made according to the directions of the Gene-P for blue eyes rather blue eyes are the result of what organisms do in the absence of the brown eye pigment. Reference to the gene for blue eyes serves as a kind of instrumental short hand with some predictive utility.

Thus far Gene-P sounds purely classical, that is, Mendelian as opposed to molecular. But a molecular entity can be treated as a Gene-P as well. BRCA1, the gene for breast cancer, is a Gene-P, as is the gene for cystic fibrosis, even though in both cases phenotypic probabilities based on pedigrees have become supplanted by probabilities based on molecular probes. What these molecular probes do is to verify that some normal DNA sequence is absent by confirming the presence of one, out of many possible, deviations from that normal sequence that has been shown to be correlated (to a greater or lesser extent) with some phenotypic abnormality. To satisfy the conditions of being a gene for breast cancer or a gene for cystic fibrosis does not entail knowledge about the biology of healthy breasts or of healthy pulmonary function, nor is it contingent upon an ability to track the causal pathway from the absence of the normal sequence resource to the complex phenomenology of these diseases. The explanatory "game" played by Gene-P is thus not confined to purely classical methods, which unfortunately has made it all the easier to conflate this meaning of the "gene" with the one I will refer to as Gene-D.

Quite unlike Gene-P, *Gene-D is defined by its molecular sequence.* A Gene-D is a developmental resource (hence the D) which in itself is *indeterminate* with respect to phenotype. To be a Gene-D is to be a transcriptional unit on a chromosome within which are contained molecular template resources. These templates typically serve in the production of various gene products—directly in the synthesis of RNA and indirectly in the synthesis of a host of related polypeptides. To be a gene for N-CAM, the so-called neural cell adhesion molecule, for example, is to contain the specific nucleic acid sequences from which any of 100 potentially different isoforms of the N-CAM protein may ultimately be derived (Zorn & Krieg 1992). Studies have shown that N-CAM molecules are (despite the name) expressed in many tissues, at different developmental stages, and in many different forms. The phenotypes of which N-CAM molecules are coconstitutive are thus highly variable, contingent upon the larger context, and not germane to the status N-CAM as a Gene-D. The expression of an embryonic form (highly sialylated, i.e., further modifed by the attachment of long chains of a negatively charged sugar) in the mature organism is associated with neural plasticity in the adult brain (Walsh & Doherty 1997) but could well have pathological consequences if expressed in other tissues—yet it would not affect the identity of the N-CAM sequence as a Gene-D. So where a Gene-P is defined strictly on the basis of its instrumental utility in predicting a phenotypic outcome and is most often based on the absence of some normal sequence, a Gene-D is a specific developmental resource defined by its specific molecular sequence and thereby by its functional template capacity; yet, it is indeterminate with respect to ultimate phenotypic outcomes.

A Gene-P allows one to speak predictively about phenotypes, but only (as Johannsen realized) in a limited number of cases and within some contextually circumscribed range of probabilities. In the absence of, for example, a full molecular-developmental understanding of the processes resulting in the pathophysiology of cystic fibrosis, it can be prognostically useful to speak of "the gene for cystic fibrosis." The normal resource, i.e., the Gene-D located at the cystic fibrosis locus for the great majority of individuals who do not have a family history of cystic fibrosis affliction, is not thereby a gene for normal pulmonary function (any more than the thousands of other genes involved in normal pulmonary function); rather, it is a member of a family of transmembrane ion-channel templates. As a developmental resource, it is one among very many that play a direct role in pulmonary development and function (as well as many other things). To speak of and direct one's attention to this gene for a transmembrane ion-conductance regulator protein is to become involved in an entirely different kind of explanatory game, i.e., that of a Gene-D (see table 1.1). There is no preformationist story to be had at this level. To study the biological role and function of this gene for a chloride channel involves locating it within all of the contexts in which it is biologically active and attempting to elucidate the causal pathways in which it is an interactant (Kerem & Kerem 1995, Jilling & Kirk 1997). And as with any developmental resource, its status with respect to cause and effect in any given interaction will be contextual and perspectival (i.e., its actions will be viewed as either the cause of something or as the result of something else, depending on how a particular inquiry is framed).

As a molecular-level developmental resource, Gene-D is ontologically on the same plane as any number of other biomolecules—proteins, RNA, oligosaccharides, and so forth—which is to say only that it warrants no causal privileging *before the fact*. Gene-P and Gene-D are distinctly different concepts, with distinctly different conditions of satisfaction for what it means to be a gene. They play distinctly different explanatory roles. There is nothing that is simultaneously both a Gene-D and a Gene-P. That the search for one can lead to the discovery of another does not change this fact. Finding the Gene-P for cystic fibrosis led to the identification of a Gene-D for a chloride-ion, conductance-channel template sequence. But the latter is not a gene for an organismic phenotype. Its explanatory value is not realized (and cannot be realized) in the form of an "as if" preformationist tool for predicting phenotypes. Rather, the explanatory value of a Gene-D is realized in an analysis of developmental and physiological interactions in which the direction and priority of causal determinations are experimentally first revealed (table 1.1).

The explanatory story in which Genes-D plays a role is *not* one of preformationism but of epigenesis. Phenotypes are achieved through the complex interactions of many factors, the role of each being contingent upon the larger context to which it also contributes. What is true for NCAM is true for the Gene-D associated with the cystic fibrosis locus, with the breast cancer (BRCA1 and BRCA2) loci, and in fact with all of the genes (Genes-D) being identified at the level of specific molecular sequence by the Human Genome Project. Gene-D, the normal molecular resource at the cystic fibrosis locus, is not a gene for healthy lungs but a genetic resource that provides template information for a transmembrane, chloride-ion channel, a protein which may be woven into cellular membranes and which plays a functional role in the transport of chloride ions into and out of the cell. Similarly, the normal resource at the breast cancer locus (BRCA1) is not a gene for healthy breasts but a template for a large and complex protein which is present in many different cell types and tissues and in many different developmental stages and which also appears to be capable of binding to DNA and influencing cell division in a context-specific way.

To study the biology of a Gene-D is to play one kind of explanatory game, an epigenetic one. To use a Gene-P, i.e., the absence of a normal genetic resource, as predictor of a phenotype is to play a different kind of explanatory game, an "as-if" preformationist one. Johannsen was not privy to Gene-D, and his injunctions do not pertain to them. He predicted that the entirety of the organism would not be decomposed into genes, and he was right. Genes-D are molecular sequences along the chromosome, not pieces of the phenotype. Genes-P are spoken of as if they were pieces of the phenotype but, as Johannsen predicted, they pertain only to a limited, and in some sense superficial, set of traits and then only for practical purposes. Now Gene-D and Gene-P can both be used responsibly within their proper domains. Genetics counselors, for example, use Gene-P. But just as the word "bank" can be properly used to mean both the side of a river and a good place to invest money, yet without implying that the side of a river is a good place to invest money, so too is the case where the word "gene" should not become

Table 1.1

Gene Concept	Examples	Explanatory Model	Ontological Status
Gene-P Defined with respect to phenotype but indeterminate with respect to DNA sequence	Gene for breast cancer Gene for blue eyes Gene for cystic fibrosis	Preformationist (instrumental)	Conceptual tool
Gene-D Defined with respect to DNA sequence but indeterminate with respect to phenotype	NCAM, actin fibronectin, tubulin 2000 kinases (28,000 other examples)	Epigenesis	Developmental resource (one kind of molecule among many)
Conflated GeneP/GeneD	—	Preformationist (constitutive)	Virus that invents its own host ("the replicator")

simultaneously invested with the meanings of both Gene-D and Gene-P. Genes are not at once both molecular sequences and pieces of the phenotype, and yet it is precisely this conflationary confusion which has buoyed up the notion of the genetic code and a blueprint that regulates its own execution.

While the template relationship that a nucleic acid sequence in DNA has to a protein (i.e., Gene-D) may be called information, and the predictability (however limited) of a Mendelian unit for the inheritance of an aberrant (or normal) phenotype may also be called "information," *these can hardly be considered the same kinds of information.* Nor certainly would it be other than shameless sleight of hand to assert that genes thereby simultaneously possess both kinds of information. Yet, in order for the claim to be redeemed that genes possess the information for making an organism, something very much like this would have to be the case. The realization of genetics understood not as a practice of instrumental reductionism but rather in the constitutive reductionist vein, would require the ability to account for the production of the phenotype on the basis of the genes. This is clearly what the rhetoric of the "genetic program," "genetic blueprint," and so forth implies. Has the discovery of the structure and mechanisms of DNA provided what classical genetics alone could not do—an explanation for the development of the phenotype? Chapter 3, following some historical considerations on the acquisition of the information metaphor in chapter 2, will further scrutinize the empirical adequacy of the idea that genes contain the information for making a phenotype.

2

The Rhetoric of Life and the Life of Rhetoric

In calling the structure of the chromosome fibers a code-script we mean that the all-penetrating mind, once conceived by Laplace, to which every causal connection lay immediately open, could tell from their structure whether the egg would develop, under suitable conditions, into a black cock or into a speckled hen, into a fly or a maize plant, a rhododendron, a beetle, a mouse or a woman . . . But the term code-script is, of course, too narrow. The chromosome structures are at the same time instrumental in bringing about the development they foreshadow. They are law-code and executive power—or, to use another simile, they are architect's plan and builder's craft—in one.
—Erwin Schrödinger, 1944

Gene-P or Gene-D

I have argued in the first chapter for a bipartite understanding of the meaning of "genes." Genes may be accorded the preformationistic status of being prior determinants of some phenotype (Gene-P) but only for a limited number of traits and only in the spirit of instrumental utility when some local benefit is to be had in doing so. The use of genetic probes for certain mutations, such as those associated with cystic fibrosis (CF), might exemplify this usage. Identification of cystic fibrosis (CF) genes, those alternative forms of the DNA template for a certain transmembrane, chloride channel protein that are implicated in the onset of CF, tells us little about the developmental physiology (including epithelial cell–microbial interactions) that actually results in CF disease, but it *has* proven to have some instrumental value in predicting an undesirable human condition. Even in the case of CF the value of the instrumental-preformationist approach tails off when one is considering the wide spectrum of different CF mutations (now up to 994[1]), the combinatorial complexity associated with correlating phenotype with the particular pairs of CF variants that could occur, and the observed failure of the same pairs to result in the same phenotypes in different individuals.

The second meaning of genes (Gene-D) refers to a segment of DNA characterized as a transcriptional unit that provides template information for some range of polypeptides but whose relationship to a phenotype is always in itself indeterminate. In this view, genes, in order to be related to an ultimate phenotype, must be situated in the dynamic developmental context and environmental milieu of an organism. As one category of internal resource, one type of molecule (or part thereof) among many, genes are not accorded any form of necessary causal privileging. I refer to this perspective, which draws on both the latest knowledge of biochemical-molecular interactions as well as that of the dynamics of complex systems, as "the new epigenetics."

The Conflation of Gene-P and Gene-D with Rhetorical Glue

My analysis of the double meaning of genes is meant to serve as a counterpoint to what I take as an attempt to have it both ways, that is, to understand genes as simultaneously both discrete segments of DNA and causally privileged determinants of phenotypic outcomes. The engagement of textual metaphors with which to characterize genes as different from other biological material, i.e., as text—program, blueprint, codescript, books of life, and so forth—has been integral to this conflationary construction. As text, and perhaps only as such, genes can be conceived as molecules and yet evade the circumstantial contingencies, the "fateful winds," which most pieces of matter find hard to resist. It is as matter-text that genes and DNA ascend to the status of sentiency and agency, as matter with its own instructions for use, and furthermore, as the user too. If such a view really exists (and indeed flourishes with increasing significance for social policy, biomedical research and development, and human self-understanding) as I suggest, then it must have emerged from somewhere and presumably (hopefully?) stand in some relationship to empirically accountable claims about the way things are. The purpose of this chapter will be to uncover at least some of the more important root sources of this idiom, to try to clarify what sort of semantic strings are attached to it, and to make salient those claims to which it could and should be held accountable.

There is an additional subtext that can be made more explicit. The idea, advanced by post-modern critics of science such as Donna Harraway and Richard Doyle, that technologies of language construction—the so-called rhetoric of science—can be as instrumental in the shaping and promulgation of a certain research program as, say, gel electrophoresis or any other central piece of instrumentation, is a view with which I have much sympathy. But while these critics often take up the tone of the indignant, if surreptitiously bemused, muckraker revealing a scandal, they abstain, as if required by some categorical imperative, from defending any truth claims at the level of the subject matter of their text.

The scandal of scandals, it would seem for these critics, is that the rhetoric of science moves inexorably, always charting its own course. Yet in seeking to clarify the claims and arguments upon which a certain rhetoric, such as that of the genetic text, could be rationally defended or undermined, I am seeking to penetrate the Teflon autonomy of the rhetorical "trope" (even if with the aid of other rhetorical tropes). My discussion in this chapter is then both a critique of the rhetoric associated with scientific research programs and a contribution to a conversation about how to criticize the rhetoric associated with scientific research.

The Self-Executing Code-Script

Erwin Schrödinger's 1944 monograph, *What is Life?*, was described by Gunther Stent (1995) as having played a mobilizing role, an *Uncle Tom's Cabin* (if you will), in rallying a new generation of physicists and chemists to the cause of working out the nature of genetic substance. It is well known that *What is Life?* was influential for both the young James Watson and the young Francis Crick (Portugal & Cohen 1977, Judson 1979). And it was one of the very few points of background commonalty between them. I suggested at the end of chapter 1 that the only alternative to recognizing the need for recontextualizing molecular genes

(Gene-D) in a renewed theory of epigenesis would have to be some theory which depicts the genotype as containing its own instructions for use.

Only if the entire process of ontogeny can be understood as progammatically prespecified in the genotype can epigenesis be relegated to a kind of epiphenomenon of the genotype, an idea which is unfortunately often promoted by naive readings of Waddington's term "epigenetics." The epigraph above reveals that Schrödinger provided just such a vision and just such a metaphor with his notion of a code-script that is at once self-executing: "architect's plan and builder's craft—in one."

Schrödinger, the Nobel prize–winning (1933) cofounder of modern quantum theory, had his attention drawn to biology during a 1935 lecture by Max Delbrück in Berlin (Judson 1979). Delbrück was the product of an elite academic family in Berlin and most highly influenced in his style and thinking by Niels Bohr. He attributed his first stimulus for thinking about biological issues to a discussion with Bohr in 1931 concerning the bearing of quantum mechanics on biology (Olby 1974).

Delbrück came to the United States in 1937 as a physicist. By 1940 he had established a research program with Luria and Hershey into the investigation of the genetics of bacterial viruses. The team became famously known as the "Phage Group" (Portugal & Cohen 1977, Judson 1979). Delbrück, one of the most influential of the physical scientists to make the move to molecular biology, was also the most prominent of those early molecular biologists whom John Kendrew (1967) classified as "informational" as opposed to "structural." The structuralists were mostly English (as was Kendrew) and generally x-ray crystallographers (as was Kendrew). The notable exception was Linus Pauling, an American at Caltech, who approached the structural analysis of biological macromolecules as a problem in quantum theory. Delbrück, guided by his goal of elucidating the physical basis of hereditary information transfer, chose to study the bacteriophage.

The "phage," as it became affectionately known, recommended itself as apparently the simplest entity that participated in some form of heritable information transfer. It consists of only two types of biological macromolecules, protein and nucleic acid (DNA). Evidence was provided by Avery and coworkers in the mid-1940s that DNA was responsible for affecting the bacteriophage-induced "transformation" of bacterial cells, yet, largely because DNA appeared to be information-poor, the Avery results were relegated to the back-burner. Proteins consist of 20 different monomeric subunits (amino acids) as compared to the merely four units of DNA, and thereby proteins appeared to Delbrück to be the more likely basis of hereditary information. For Delbrück, the leader of the informationist camp of the physicists-cum-molecular biologists, the heuristics of "information" pointed *away* from DNA and the "path to the double helix."

Delbrück's influence on Schrödinger can be interpreted in different ways and has been the subject of some differences of opinion. Robert Olby (1974), mostly taking issue with Gunther Stent's characterization of the so-called romantic phase of molecular biology, has argued that Schrödinger was not seeking the discovery of new fundamental laws of physics in biology and/or the realization of Bohr's principle of complementarity in the life sciences, but rather he was attempting to show that the quantum theory of the chemical bond could account for the requisite stability of biological order in a way which classical statistical physics could not:

He [Schrödinger] could demonstrate that even if genes were not chemical molecules the physicist could still allow them to have stability and mutability, because of quantum mechanics. Now a physicist reading this book could get excited about genetics. Schrödinger made the facts of genetics meaningful to the physicist. He did *not* offer his readers the bait of a fresh mutually exclusive complementarity relationship, as did Bohr. He offered them "other laws" to be sure, but not of the kind Bohr envisaged. They would be related to known physical laws, just as the laws of electrodynamics were related to the more general laws of physics (Olby 1974).

An examination of Schrödinger's text reveals that Olby is correct in his appraisal of Schrödinger's motivation for writing the book. Olby (1974), however, does not adequately analyze Schrödinger's warrant for moving from his quantum mechanical arguments to his promotion of the idea of a hereditary code-script, yet he celebrates Schrödinger's "concept of an hereditary code-script" as that which "we can see in retrospect as the most positive and influential aspect of this little book." Influential? Without a doubt. But positive? *Why?*

Schrödinger's central concern is with that most perennial (and perennially elusive) topic of those who would presume to "explain" life—

i.e., the acquisition, retention, and propagation of organized form. Schrödinger has set himself the task of showing that there has been a certain special problem and that now the problem can be solved. The problem for Schrödinger is that statistical physics cannot account for the stability which genetics has shown must be invested in only a single or perhaps two copies per cell capable of stability over numerous generations. His favorite example is the "Hapsburg lip," a Mendelian trait (autosomal dominant) observed to be faithfully reproduced in its passage through many generations of Hapsburgs. Schrödinger's solution to this problem is to be found in the new explanatory insights into the stability of the solid state which quantum mechanics claims to provide and in the specific form of the 1926–1927 Heitler-London theory of the chemical bond. Schrödinger begins by presupposing the kind of "constitutive" genetic preformationism I have already attempted to criticize. What he proceeds to do is to make the question of inheritance interesting to physicists by recouching it in terms of the physics of stability and predictability. He follows Delbrück in uniting the idea of genetic transmission with that of information given a physical meaning by quantum mechanics. The product of this marriage is the concept of the "hereditary code-script."

Although he begins with an uncritical acceptance of genetic reductionism, Schrödinger proceeds to offer an independent argument for the need for a genetic code-script. It is this argument which has a special appeal to physical scientists. Schrödinger's case for the hereditary code-script is based on distinguishing between "order-from-disorder" and "order-from-order," with the former being the predominant legacy of statistical physics and the latter the result of new insights from quantum theory that could meet the challenges to biology which the former framework could not. Schrödinger begins with a naive notion (but perhaps justifiable for his time) of the cell as a disorganized bag of atoms and argues his way to the need for a solid-state "aperiodic crystal" to serve as that bedrock of order, continuity, and heroic resistance to entropy which makes life possible. Those who have come since and who continue to sing the praises of the hereditary code-script have failed to examine these accompanying presuppositions in light of the empirical findings that have since accrued.

Schrödinger tells us that natural order, as hitherto described by physical laws, was based on atomic statistics and so only approximate. Lawful precision—as demonstrated by the examples of paramagnetism, Brownian motion, and diffusion—is predicated on large numbers of interacting atoms. The relationship of the number of units in the system to the predictability of the system is given by the "square root of n rule," with n equal to the number of interacting units in the system:

The laws of physics and chemistry are inaccurate within a probable relative error of the order of $1/\sqrt{n}$, where n is the number of molecules that co-operate to bring about that law—to produce its validity within such regions of space or time (or both) that matter, for some considerations or for some particular experiment (Schrödinger, p. 19).

A system with only 100 molecules would see a relative error of 0.1 while a system of 1 million molecules would see only a relative error of 0.001. Schrödinger then goes on to draw upon the evidence of classical genetics in order to argue that the stability in living systems entails the stability of small numbers of units. A gene, for example, appears to be composed of approximately 1000 atoms, as was estimated by Delbrück, using x-ray induced mutation data. The overriding point is that biological stability simply cannot be accounted for on the basis of those laws which statistical physics has to offer for explaining the stability of molecular systems.

Having established the problem, Schrödinger offers to provide the solution. The solution, he believes, is to be found in quantum mechanics, and in particular in the 1926–1927 Heitler-London theory of the chemical bond. First and foremost for Schrödinger, quantum mechanism introduced a break with the ontology of continuity:

The great revelation of quantum theory was that features of discreteness were discovered in the Book of Nature, in a context in which anything other than continuity seemed to be absurd according to the views held until then (p. 51).

Why this should be important is not hard to imagine. In a world of unbroken continuities, predictability will vary along a continuum as described by the $1/\sqrt{n}$ law. Configurations consisting of small numbers of atoms will always be relatively unstable. In a world of discontinuities, however, critical thresholds can emerge which can account for the preservation of stability within some sub-threshold regime.

Atomic configurations, as described by quantum mechanics, do not vary along a continuum but rather are limited to some set of allowed states. And of the allowed states there may be a lowest energy or most stable configuration. Such a configuration will then be separated from the next most stable configuration by a discrete difference in energy that would be required for a transition of states to occur. Now if all configurations were "allowed," then energy differences would presumably run along a continuum and fluctuations would be ongoing. But with allowed configurations being constrained at the most fundamental level, discrete differences in the energy of allowed states can constitute relative barriers and thus provide for relative stability.

Following Delbrück, Schrödinger formalizes this relationship as follows. If W is the energy difference between two (allowed) molecular configurations, then stability will be a measure of the ratio of W to the average heat energy of the system, which is given by kT, where k is known as Boltzmann's constant and derived from the average kinetic energy of a gas atom at room temperature, and T is the absolute temperature. The "time of expectation" for a transition of states to occur, represented by t, which is a measure of the probability of enough energy gathering by chance in one part of a system to effect a transition, is given as an exponential function of the ratio of $W:kT$ and is thus highly sensitive to changes in this ratio. With τ representing the time in seconds for a molecular vibration (10^{-13} or 10^{-14} s) to occur, the full expression is $t = \tau e^{W/kT}$. The sensitivity of this equation to changes in the ratio of $W:kT$ is such that while the time of expectation for a transition to occur with a ratio of $W = 30\,kT$ is only one-tenth of a second, t goes up to 16 months when $W = 50 \times kT$ and up to 30,000 years when $W = 60 \times kT$. To this picture Schrödinger added two amendments, which will be familiar to anyone acquainted with contemporary chemistry.

The first is that the next highest energy level does not actually entail molecular rearrangements:

The lowest level is followed by a crowded series of levels which do not involve any appreciable change in the configuration as whole, but only corresponds to those small vibrations among the atoms which we have mentioned (p. 56).

These vibrational states are also quantized (and any given molecular configuration will be compatible with some range of vibrational states). The next amendment pertains to the atomic mechanisms involved in an actual mutational event. A mutation, caused perhaps by radiation, may entail a rearrangement of the molecular configuration such as to produce an isomer. "Isomers" are molecules that are described by the same chemical formula but have different spatial arrangments. Now two isomers may be fairly close with respect to their lowest energy state, and yet the transition from one to another may still be highly constrained. The reason for this constraint is that the mechanism used in getting from one isomer to another is an intermediate configuration which is of a much higher energy level than either of the isomers. In other words there is not a direct mechanical path from one isomer to another but only a path that entails an intermediate state which is at a significantly higher energy than either of the two stable isomers. There is typically thus a significant "energy of activation" which intervenes between two otherwise fairly stable, low-energy configurations.

The physics of the chemical bond described by the Heitler-London theory pertains equally to molecules, solids, and crystals. From the quantum-mechanical point of view these terms all represent one and the same state of matter. The notion of Schrödinger's celebrated aperiodic crystal is simply that of a molecule which enjoys this stability and is sufficiently lengthy and sufficiently heterogeneous in its composition to be the putative bearer of coded information which allows for the sustenance and reproduction of organized life-forms. Why organized life-forms must be dependent upon coded information is never explicitly addressed but can be inferred from Schrödinger's reasoning. His view requires, among other things, the assumption that no other aspects of the cell are capable of 'standing on their own' with respect to preserving high levels of order. If the bulk of the cell is not based on some form of solid-state organization and yet is dependent on it, then that order retained in a relatively inert solid-state form—being at a kind of remove from the dynamics of metabolizing life processes—must exist as something like a representation of those dynamics, embedded perhaps, in some form of code. Schrödinger approaches this kind of vision in the following sentence:

An organism's astonishing gift of concentrating a "stream of order" on itself and thus escaping the decay into atomic chaos—of "drinking orderliness" from a suitable environment—seems to be connected with the presence of the "aperiodic solids," the chromosome molecules, which doubtless represent the highest degree of well-ordered atomic association we know of—much higher than the ordinary periodic crystal—in virtue of the individual role every atom and every radical is playing here (p. 82).

Schrödinger did not attempt to adumbrate any of the laws or mechanisms by means of which the order of chromosomes serves to dictate and direct the dynamics of an organism. Rather he holds this out for future biochemical physiologists to work through. But it is the very idea that there are such new laws to be found which Schrödinger asserts is the principal motivation for writing his book *What Is Life?*

What is novel for the physicist about the living organism is that its exquisite stability and predictability is the result of an "order-from-order" process. This new order-from-order is not being put forward as evidence for new fundamental laws of physics nor is it seen to contravene any of the established laws. Rather, Schrödinger foresees the finding of new higher-level laws or principles that explain the ability of living systems to parlay high levels of order between the chemically stable but metabolically inert aperiodic crystal and the growing and metabolizing, but entropically vulnerable, apparatus of the cell and organism. He likens this to the example of the spring-based clock. The clock is made of real physical stuff and so cannot in principle escape the possible consequences of thermal fluctuations. No matter how improbable, the possibility that the clock could suddenly go wrong (in reverse even) always remains in the background. The means by which the clock usually keeps proper time are the order-from-order principles of the clockwork.

Drawing on Planck's distinction between "dynamical" and "statistical" laws, Schrödinger asserts that "Clockworks are capable of functioning 'dynamically', because they are built of solids, which are kept in shape by Heitler-London forces, strong enough to elude the disorderly tendency of heat motion at ordinary temperature" (p. 91).

Dynamical processes are the achievement of some form of organization which, for all intents and purposes, behave as if they were at absolute zero; that is, they behave as if random thermal motion were not a factor. The working clock at room temperature functions as if it were at absolute zero inasmuch as thermal fluctuations are rendered irrelevant to the macroscopic dynamics of the system. Life too is an order-from-order system, but whereas the order-from-order principle of the clock is known to the clockmaker, the order-from-order principles of life are yet to be determined. This order, in the first instance for Schrödinger, must be the result of a solid—the aperiodic crystal—being withdrawn from the disorder of heat motion. Life must then acquire its ability to elude or overstep the degradative forces of heat motion through what, in effect, will come to be called a "translation" of the order of the aperiodic crystal to that of the dynamics of the cell (and beyond).

Schrödinger's vision is powerful and recognizably influential but it is based on certain hypothetical (presumably empirical) assumptions that should not escape our notice. For Schrödinger the "game of life" is about evading the entropic decay inevitably associated with the statistical interactions of thermal physics. The idea of dynamic self-organization arising out of statistical interactions—that is, order arising from out of disorder, the very battle cry of the apostles of nonequilibrium thermodynamics—is exactly what Schrödinger denies. Indeed, it is just because the seat of biotic order must thereby be secured through its removal from the heat flux of the cell and, owing to its inertness, its exclusion from the biotic dynamics of the cell that its characterization as a code-script presents itself as apropos. It is exactly because the aperiodic crystal is secured and sequestered from the hurley-burley of chemical dynamics, in this more ethereal notion of a code, that the idea of a "translation" from the code of the crystal to the chemistry of the cell would seem to follow.[2] Now beyond the holus-bolus assumption of equilibrium thermodynamics (which is bedrock for Schrödinger and will be discussed later, especially in relation to Stuart Kauffman's work) there are also more discrete assumptions and/or assertions to which Schrödinger's view can be held empirically accountable.

Is the aperiodic crystal of the chromosome really the unique bastion of molecular order in the cell? Does all expression of ongoing order truly derive from it? If not, then what are the other sources of ongoing order and what is the nature of the relationships between them? Does the aperiodic crystal itself depend for its stability on other sources of cellular order, or is it truly foundational? These questions must be answerable in a fairly unequivocal way for Schrödinger's strong characterization of the self-executing code-script to be ultimately justified. Chapter 3 will address these matters in detail.

Getting a Grasp

In his discussion of Schrodinger's book and its impact on biology, rhetorician and post-modern science critic Richard Doyle (1997) examined a progression of metaphors in Schrodinger's text that he characterized as "slippage." The organism for Schrödinger becomes replaced by its phenotypic "pattern," and the phenotypic pattern, in turn, is replaced by its code-script. Unlike Doyle, I have endeavored to reconstruct from Schrödinger's text an argument on behalf of the hereditary code-script metaphor that is rationally compelling given the adequacy of certain associated assumptions and hypotheses. I would thus prefer to target the charge of slippage (if slippage can be the stuff of accusations) not so much at Schrödinger as at those who have subsequently embraced and/or celebrated the self-executing code-script metaphor without considering under what set of assumptions it is warranted to do so. So stated, Olby (and so many others) become the culpable parties. To attempt to put forward such a distinction is certainly to take a step in the direction of making that which I referred to earlier as a subtext more explicit. I am under no illusions about the possibility of anchoring some piece of knowledge in rhetoric-free prose, but I will not thereby grant that nothing remains but a battle of the tropes. What "relativizes" the power of the phrase to a larger context of reasons and thus mitigates its autonomy, is the teleological framework of a problem-solving orientation. Slippage is an evocative and appropriate description of much of the discourse dynamics that have led to shallow and confused talk about genes as texts. What has slipped is the metaphor or trope, such as code-script, from out of a context in which the conditions of its warrant can be judged—i.e., in which it can be held accountable—to one in which it appears to be self-sufficient. What gives slippage its normative bite is the sense of something dropping out its proper place, the place in which it participates in some form of normal functioning, where it allows "things" to work. The normativeness of slippage presupposes a teleological framework.

An example of slippage can be found within Schrödinger's text. One area in which Schrödinger already anticipated a challenge to his position is that associated with multicellularity. Schrödinger had asserted that organisms are dependent on single or double copies of their aperiodic crystals, yet multicellular organisms have single or double copies in each of very many (10^{14} in a grown mammal, he suggests) cells. From the point of view of a whole metacellular organism there is a question of order-from-order versus order-from-disorder on an entirely different level. Cell biologists subsequent to Schrödinger have in fact routinely found a great deal of cell-to-cell heterogeneity, which raises the specter of order arising from out of disorder at a higher level of analysis, i.e., that on the order of a unified organism and its disparate and heterogeneous cellular constituents. It would appear that organismal order may then require a different kind of explanation, one which cannot be physically accounted for on the basis of the solid-state structure of chromosomes, because a whole realm of disorder lies between the chromosome and the organism as a whole. Schrödinger's comment, more "poetic than scientific" by his own admission, serves only to place what was and continues to be an intriguing problem even further from the reach of reason:

Since we know the power this tiny central office has in the isolated cell, do they not resemble stations of local government dispersed through the body, communicating with each other with great ease, thanks to the code that is common to all of them? (p. 84)

In this case, Schrödinger does not provide any independent arguments at all but rather insinuates the accepted presence of one metaphor, that of a "tiny central office," and builds on it by a kind of associative logic of the image, a notion of the constituent body, governed by an interlocking network of local offices which communicate with one another by means of a code that we should now take as a given, not as a hypothesis. Schrödinger's trope, which ascribes governing agency to the inert aperiodic crystal by metaphorical inference, does not readily lend itself to inspiring new research programs concerned with cell-to-organism achievements of order, but it has nonetheless proven to be Teflon-coated. The rhetorical progeny of Schrödinger's metaphor *slip* into various public media expositions of heredity, carrying the image of a detached, and thus preformed, genetic determinism with them.

Gamow's Translation

In discussing Schrödinger's solution to the problem of biological organization, the notion of translation has already been broached. By placing the locus of biological order on the side of a code-script embedded in the entropy-resistant calm of the solid state, Schrödinger's model begged for some transitional principles with which to bridge the chasm between code-script calm and the moving parts of the cell. Translation as such does not become an important theme until after Watson and Crick characterize the structure of DNA, but then it does it so very quickly. The impact that cybernetics, information theory, and linguistics had during the postwar environment of the early 1950s on the reception and interpretation of the new molecular biology has become the topic of recent scholarship (Kay 2000). George Gamow, a Russian émigré physicist, science popularizer, and military strategist, responded immediately to the Watson and Crick paper by approaching the relationship between DNA structure and protein synthesis as a cryptanalytic problem (Kay 2000). It was as a problem in the cryptanalytic breaking of a code that the transition between code-script and metabolism implicit in Schrödinger's book became explicitly about *translation*. Cryptanalysis was not new but it was in the midst of being transformed by Shannon's information theory and by the rise of the computer. As Kay has perceptively pointed out:

[It was] the simultaneous transportation of cybernetic and informational representations into both linguistics and molecular biology in the 1950s that propelled the striking analogies between the two fields. As in other disciplines, it is through the circulation of the information discourse that their objects of study were (separately) reconfigured anew, and then emerged, not entirely surprisingly, with some parallel features. And it is this simultaneous dematerialization of both language and life that soon formed the conditions of possibility for envisioning the word (information of the DNA sequence) as the origin of self-organization, the ontological unit of life and evolution (Kay 2000).

Gamow's first response to the Watson and Crick double helix was to approach the Navy's Bureau of Ordnance, which had deciphered and broken Japanese code, with lists of amino acids and DNA bases to put into its deciphering computer. Gamow reported that "after two weeks they informed me there is no solution" (in Kay 2000). Gamow (1954) quickly offered a solution of his own, known as "the diamond code," which entailed a direct physical translation between DNA and amino acids. The diamond code consisted of all the unique arrangements of the four DNA bases organized into a diamond shape. As it happens there are exactly 20 diamond-shaped configurations that correspond numerically with the fact of 20 amino acids. This became known as the "magic twenty." Gamow's idea was that each diamond configuration, defined by the arrangement of four bases, one at each of the vortices, resulted in 20 unique rhomb-shaped holes into which each of the amino acids would fit. Gamow's model described an unmediated direct physical relationship between DNA and protein. Indeed, had Gamow's model held up, the notion of a direct translation between DNA and protein sequence would have to be granted a certain literal legitimacy. That Gamow's diamond model could not be empirically supported is of prime importance in critically examining the rhetorical trajectory of molecular biology.

Gamow's diamond quickly met with criticism from the likes of Linus Pauling, Erwin Chargaff, Francis Crick, and Martin Yčas, whose thinking relied upon, amongst other things, inferences made on the basis of Fred Sanger's newly sequenced insulin. As a visiting professor at Berkeley in the fall of 1954, Gamow elicited the interest of Caltech luminaries Delbrück, Alex Rich, and Richard Feynman with his desire to pursue the coding problem. The role of RNA as an intermediate in protein synthesis had gained favor at Caltech in the context of an atmosphere in which there was already an appetite for using the most sophisticated mathematical methods and computer resources available and approaching the problem as one akin to breaking enemy code. Gamow formalized the group as the "RNA Tie Club," consisting of 20 members, each issued a diagrammed tie, designed by Gamow, representing one of the amino acids (Kay 2000).

Approaching the *coding problem* as one of analyzing a language meant looking for the kinds of restrictions which are characteristic of natural languages. The distribution of letters in words are not random but follow rules or patterns that pertain jointly to all, or most, of the words in a language. Many such restrictive (overlapping) codes were proposed by some of the preeminent theorists of the day (von Neumann, Wiener, Gamow, Rich, Crick, and Delbrück) using the Los Alamos Maniac computer for assistance. If codes were fully overlapping, then, for example, a sequence of ABCD would code for two amino acids, one based on ABC and the other on BCD. Being fully overlapping avoids any problem of punctuation because every base is the beginning of a new codon, but it will also place restrictions on which amino acids can precede or succeed any particular amino acid as each codon must share two-thirds of its sequence with the codons that come before and after it. Not every amino acid can precede or succeed any other amino acid. Models of overlapping codes with their attendant restrictions could be examined theoretically and with reference to known sequences of amino acids and DNA:

The results were negative. In spite of the strong inter-symbol restrictions in the proposed codes, both the artificial sequences constructed based upon them and the naturally occurring sequences produced a random amino acid distribution. Rather than question their guiding premise that the code was overlapping, or more fundamentally, that the scheme was in fact, a language like code, the team inferred that the method employed was not sensitive enough (Kay 2000).

The efforts of the RNA Tie Club continued for 5 years, producing many hypothetical codes but little biological accomplishment. While the heuristics of "nucleic-acid sequence as language" did not prove to be biologically fruitful, it left an enduring legacy at the level of the speech-style used to describe and conceptualize molecular-level biology.

Rhetorically fashioned by information theory, DNA emerged as a close cousin of language and encryption. The random distribution of elements in both DNA and protein contravene the expectations for a language to be overlapping and restrictive, yet the insinuation of semantic properties in the DNA/RNA sequence continued to accrue. In approaching the challenge of a nonoverlapping code it was pointed out that any nonoverlapping code would have to account for how at any given locus "a base sequence read one way makes sense, and read the other way makes nonsense" (Yčas 1969). DNA strands, to this day, are distinguished as *sense* and *nonsense*, and not only does template-based polymer synthesis count as *reading* but also a complex array of enzyme-based *proofreading* mechanisms have become part of the canon as well. Another enduring piece of language metaphor was introduced by way of an empirically unsuccessful attempt to resolve the challenge of a nonoverlapping code and the absence of punctuation marks.

Given the fact of an ongoing sequence of nucleic acid bases which are *read* three at a time, how is it possible for them to be read in only one register, i.e., how is the start and stop of each triplet distinguished? Crick, Orgel, and Griffith offered their scheme for a comma-free code in 1956. The solution they proposed illustrates the influence of the linguistic metaphor. Their solution consists of partitioning all the possible codons (sequences of three) into those that have "meaning" and those that don't. The key to this partition is to locate all and only those codons that when taken together can be juxtaposed without creating the possibility of a meaningful codon existing by overlap. So, for example, for ABC and DAB to meaningful, it would require that BCD and CDA, as well as ABA and BAB, could not be. All the meaningful codons would then constitute a "dictionary" by their terms. In this way a sequence would be *read* univocally and without the need for punctuation. Although this did not prove to be the solution to the comma-free code problem, the rhetoric of textuality grew further and prospered.

The coding problem, that is, the basis by which RNA/DNA serves as a specific resource in the biosynthesis of proteins, was solved entirely outside of the organizational associations of Caltech and the RNA Tie Club, as well as the whole of the information-theoretical apparatus. Marshall Nirenberg and Heinrich Matthaei at the National Institutes of Health solved it empirically using methods of wet biochemistry. But despite the fruitlessness of the efforts of Gamow and his theoretically minded collaborators, their stylization of the relationship between DNA and everything else in terms of linguistic and crytographic metaphors was successful.

Listening to the Words

From Schrödinger's code-script to the fashioning of DNA in terms of *translation*, *sense* and *nonsense*, *reading* and *reading frames*, *meaning*, *dictionaries*, *libraries*, and the like, the imputation of semantic content has followed its own logic and served to set the stage for subsequent emerging perceptions and interpretations in the discipline. I have suggested that it would have made a difference if something like Gamow's diamond code—in which DNA and proteins would have interfaced in close physical register—had proven to be warranted. Translation then, in its immediate specification, might have become innocuous in its victory, progressively taking on more of the character of a technical term denuded of its broader resonances. But as the characterization of molecular interactions has evidenced greater complexity and multidirectionality of effect, the impact, if anything, on the semanticizing idiom has been to cut it more slack, to allow it the opportunity to veer off in its own direction with but a fluttering empirical tether. If the rhetorical dimension of the history of the gene has to some extent charted its own course, then an attempt such as Doyle's to examine the "rhetorical transformations of the life sciences" is warranted.

As in the cases of a gestalt-switch experience, perceiving the duck or the rabbit, for example, the use of a metaphor to structure a perceptual field organizes the relationship between foreground and background. Doyle is especially concerned with making manifest what assumptions are marshaled together as the necessary background correlate of structuring a field by way of a certain rhetorical trope. This marshaling together of background assumptions Doyle, borrowing from Deleuze and Guattari, attributes to "order-words." He targets translation as a pivotal move, an "order-word" in the rhetorical transformation of the life sciences he seeks to characterize. Doyle calls on the assistance of the philosophers Martin Heidegger and Walter Benjamin to "think the unthought" of translation as an organizing theme. From Heidegger's essay on the "The Age of the World Picture," (1977) Doyle borrows the idea that a scientific research program opens up a space for investigation by "sketching out in advance" the nature of the terrain, the basic ground plan, upon which anything that can count as an object of interest can show up. Doyle refers to this as "an extrascientific, ontological gambit." Just what is the nature of the ground plan that Doyle alleges molecular biology came to lay out in advance? He takes his cues in this regard from Walter Benjamin's critique of the idea of translation as it is expressed in an essay entitled "The Task of the Translator" (1969) (which Benjamin wrote as an introduction to his translation of Baudelaire). Benjamin strives to supplant the understanding of translation as an exercise in providing a bridge between two stagnant bodies with the idea that translation becomes part of the ongoing life history of both languages—or perhaps of "languaging":

Translation is so far removed from being the sterile equation of two dead languages that of all literary forms it is the one charged with the special mission of watching over the maturing process of the original language and the birth pangs of its own (Benjamin 1969, p. 73).

Translation for Benjamin is a testament to the fundamental kinship of all languages which spring from a common source and mature and age and risk senility and yet which may, by the hand of the translator, refresh and renew one another:

Translation thus ultimately serves the purpose of expressing the central reciprocal relationship between languages . . . As for the posited central kinship of languages, it is marked by a distinctive convergence. Languages are not strangers to one another, but are, *a priori* and apart from all historical relationships, interrelated in what they want to express (p. 72).

What Benjamin is gesturing toward is a primordiality that underlies the intentions of each and every particular language and becomes transiently exposed during moments when historical languages are being juxtaposed. To probe deeply enough into the nature of translation for Benjamin is to embark on the path to "pure" language, the ground of all languages, inaccessible to mortals yet the ontological substratum whose light is revealed in successful translation:

Rather, all suprahistorical kinship of languages rests in the intention underlying each language as a whole—an intention, however, which no single language can attain by itself but which is realized only by the totality of their intentions supplementing each other: pure language (p. 74).

In translation the original rises into a higher and purer linguistic air, as it were. It cannot live there permanently, to be sure, and it certainly does not reach it in its entirety. Yet, in a singularly impressive manner, at least it points the way to this region: the predestined, hitherto inaccessible realm of reconciliation and fulfillment of languages (p. 75).

A real translation is transparent; it does not cover the original, does not block its light, but allows the pure language, as though reinforced by its own medium, to shine upon the original all the more fully (p. 79).

It is the task of the translator to release in his own language that pure language which is under the spell of another, to liberate the language imprisoned in a work in his re-creation of that work (p. 80).

Melding Heidegger's and Benjamin's contributions together, the onto-logical gambit which Doyle imputes to Gamow's molecular translation research program is one in which the presumption of an underlying pure language, a kind of Pythagorean "positivist mysticism," the unthought guarantor of translatability, is that which is sketched out in advance. In this sense the trope of translation, while an expression of an interest in explaining the basis of the transmission of heritable traits, is also an explanation for the very interest in breaking genetic codes.

Talking Back

If the primordial substratum, that which is ontologically laid out in advance, is that of hidden geometrical-cryptographic relationships, then puzzle-code solving follows as the way to go in attacking it. Of course just what it is that is laid out in advance is neither uniquely determined nor secured once and for all. Doyle's hermeneutics of translation is (or should be) offered in a speculative and suggestive vein. It provides an alternative to the concept of simple reductionism for elucidating the path by which life becomes increasingly assimilated into textual metaphors. Doyle's rhetorical analysis offers a much richer purview of how life "shows up" in the light of some new rhetorical regime which then colors the intuitions that are drawn on for grappling with the next set of prob-lems. But the projected ground plans are themselves subject to destabi-lization and transmutation.

Perhaps if Gamow's diamond code had met with empirical support, then some form of Pythagorean positivism would have become stabilized as the ground of translatability. But the diamond code did not carry the day, and the ground of translatability became increasingly linguistic in character. Translation is variably seen as a movement from digital to analog (von Neumann), from DNA digits to protein words, and from words (DNA) to deeds (enzymes and, thus, metabolism). Even Doyle acknowledges that each rhetorical stand opens up new explana-tory gaps; evidently the force of life's empirical being cannot just be talked away:

The gap and border between DNA and protein, numbers and words, codes and organisms is both the site of imprecision and the site of metaphorical intervention. The problem of "translating" life is one possible way of deciding on and effacing the border between textuality and vitality, a translation that appears within an *episteme* in which "Life becomes one object of knowledge among others," an object in and of language. It is a solution made possible by the simultaneous rhetorical displacement of the question of the organism and its return, a haunting trace of life that stalks the borders between codes and bodies. This imprecision of life seems to provoke a rhetorical crisis; each trope we deploy—code-script, translation, program—seems to provoke different conceptual blind spots, oversights that then render any account of living systems inadequate, imprecise (Doyle, p. 59).

Doyle's objective has been to analyze the role of "rhetorical software," with its concomitant restructuring of our ambient background presuppositions, in transforming the meaning of life. The "linguistification" of life bears for Doyle an unsettling resemblance to the decentering move of post-structuralism which announced that it was not man who speaks language but rather language which speaks man. Doyle finds in the totalizing language of the language-of-the-gene an ironic (but perhaps not) echo of the deconstructionist project with which he identifies his own intellectual patrimony:

The conflation of what life "is" with the "action" of a configuration of molecules conventionally represented by an alphabet of "ATCG" produced an almost vulgarly literal translation of Jacques Derrida's famous remark, "*Il n'y pas hors du texte.*" Literally, the rhetoric of molecular biology implied, there is no outside of the genetic text. No body, no environment, no outside that could threaten the sovereignty of DNA (p. 109).

But what is Doyle's own standpoint as an apparent critic of the rhetoric of the gene? What is the tenor of his criticism? Is he primed to hear only the most totalizing intonations by having already presupposed the uncontested agency of the language of the language-of-life and thus having relegated himself to the status of a cynical Cassandra? Does Doyle's methodological standpoint obviate the very possibility of conducting criticism with an emancipatory intent? One wonders.

The trope of language as a medium for modeling or allegorizing life may well bring with it a penumbra of meaning and association, but it does not fix or easily circumscribe what those meanings and associations can be. Construed as language, DNA could just as well (and I would suggest better) be analyzed as context-dependent "utterance" than as some form of primordial Holy Writ. Now even if, as Doyle has intimated, it is the latter image that has tacitly tended to hold sway, what is it, if anything, that would insulate it from criticism? A dialogical philosophy of language could help itself to the code-alphabet metaphors and yet "discover" biological meaning, i.e., an adapted phenotype, to be always an achievement reached via dynamic, developmental interactions and thus inevitably realized at the end of the day and never before the break of dawn. Biological agency, from this angle, is rediscovered in *process* and at many levels of context, linguistic metaphors notwithstanding.

A source of theoretical-rhetorical resources for reconstructing the metaphorics of life, in the idiom of dialogue, can be found in Bahktin (and his circle) as well as in other quarters. Consider the following quote from *Marxism and the Philosophy of Language* (Volosinov 1973), which, if we analogize a gene with a word, as publicists for the Human Genome Project have done on a regular basis, serves precisely to undermine vectoral unidirectionism (all causality emanating outward from the genes as the "deep text which underlies all else"):

Context of usage for one and the same word often contrast with one another. The classical instance of such contrasting contexts of usage for one and the same word is found in dialogue. In the alternating lines of a dialogue, the same word may figure in two mutually clashing contexts. Of course, dialogue is only the most graphic and obvious instance of varidirectional contexts. Actually, any real utterance, in one way or another or to one degree or another, makes a statement of agreement with or a negation of something. Contexts do not stand side by side in a row, as if unaware of one another, but are in a state of constant tension, or incessant interaction and conflict.

The context, in this perspective, determines the significance of the word, not vice versa. Contexts, in a biological vein, would be found at the many levels of structured, dynamic systems that are always in some relationship to other such structured, dynamic systems, and/or a complex environmental ambience.

That even the simplest free living cell is capable of considerable adaptive plasticity—i.e., successful participation in highly variant "dialogical contexts"—speaks well to the explanatory potential of depicting genes as words whose significance is context-dependent. The contrast between textualized genes as deep-seated primordial meaning and textualized genes as context-dependent utterances in particular "language games" corresponds well with the distinction between conflationary genes

(Gene-P/Gene-D) and Gene-D (see table 1.1). The critique of gene conflationism offered at the end of chapter 1 can, and should, be a vehicle for shifting rhetorical gears. The understanding of Gene-D as a context-dependent molecular resource is nicely complemented by the "metaphorics" of the dialogic construction of meaning in context, and such metaphorics can indeed be productive in elicting intuitions that biologists will be able to realize in new experimental designs. Doyle touched on the idea of giving the language metaphor a dialogic, or at least contextual, turn in his reference to the work of theoretician Howard Pattee, but he left it on the sidelines.

What Doyle, it seems, will not do is put forward his own claims based on an appeal to empirical evidence. Now surely one can and should acknowledge that there is no pure or original access to empirical evidence and that our practices—theoretical, experimental, rhetorical, and so on—have an impact on what shows up as "empirical." Yet, as I hope to demonstrate, none of these are interpretatively unequivocal. One can enter the fray, albeit critically, at the level of putting forward positions about biology in the mode of empirically accountable, problem solving–oriented claims and still make reflectively explicit the rhetorical resources that are mobilized in doing so.

3

A Critique of Pure (Genetic) Information

Omnis cellula a cellula.
—Rudolph Virchow, 1855

At its heart, the debate centres on the extent to which the sources of order in molecular biology lie predominantly in the stable bond structures of molecules, Schrödinger's main claim, or in the collective dynamics of a system of such molecules. Schrödinger emphasized, correctly, the critical role played by quantum mechanics, molecular stability, and the possibility of a microcode directing ontogeny. Conversely, I suspect that the ultimate sources of self-reproduction and the stability requisite for heritable variation, development and evolution, while requiring the stability of organic molecules, may also require emergent ordered properties in the collective behaviour of complex, non-equilibrium chemical reaction system . . . The formation of large aperiodic solids carrying a microcode, order from order, may be neither necessary nor sufficient for the emergence and evolution of life. In contrast, certain kinds of stable collective dynamics may be both necessary and sufficient for life.
—Stuart Kauffman, 1995

The phenotype is the result of ontogenetic development. This holds true also at the molecular level, because molecular biological processes take place within the organism. In ontogenesis, genetic and non-genetic factors interact in producing successive states, each of which is the prerequisite, and determines the conditions, for the next one to follow. In this interplay, genes are a necessary, but not a sufficient, component. The structures already present, gradients, threshold values, positional relationships, and conditions of the internal milieu, are equally essential. Thus even monofactorial traits can be considered to be of multifactorial causation, and the varying borderline conditions that arise during development add to the complexity. From this standpoint, it is not expected that a mutation has a consistent phenotypic outcome, and the phenotype-genotype relationship may be irregular.
—U. Wolf, 1995

Taking Schrödinger Seriously

Schrödinger's hereditary code-script metaphor is as good a place as any to locate the inception of the gene-as-information formulation, but the power of Schrödinger's text to mobilize the group of scientists who went on to constitute the new science of molecular biology was not based upon the seductions of Schrödinger's rhetoric alone. By approaching the physical basis of biological inheritance in terms of thermodynamics, Schrödinger elicited the strong interest of physical scientists. The claim that the kind of stability required of living systems is *uniquely* secured and maintained as tacit information heterogeneously embedded in the solid state structure of the aperiodic crystal might have turned out to be true. If it had been shown to be warranted, then the metaphor of the hereditary code-script could justifiably be celebrated as nothing but salutary. In criticizing the rhetoric of genetic information, it does matter whether the Schrödinger's formulation is empirically robust.

Picking up, in effect, where Schrödinger left off, the effort here will be to demonstrate and argue the following: *Neither DNA nor any other aperiodic crystal constitutes a unique repository of heritable stability in the cell; in addition, the chemistry of the solid state does not constitute either a unique or even an ontologically or causally privileged basis for explaining the existence and continuity of order in the living world.*

I will approach this task through characterizing and discussing several fundamental, non-solid-state sources of stable and heritable biological order which are not reducible to the dictates of the genome. Following the categorizations of Jablonka and Lamb (1995), these will now be grouped into three "epigenetic inheritance systems" (EISs): first, organizational structure; second, steady-state dynamics; and third, chromosome marking. While it is not my intent to imply that all of the "developmental resources" (Oyama 1986, Griffiths & Gray 1994) instrumental in determining the course of an ontogeny are encased in the fertilized egg, for present purposes I will confine my focus to that which is materially transmitted between generations in the most immediate sense.

By "organizational structure" I refer to those membranous partitions and proteinaceous matrices which constitute and dictate the three-dimensional form of the cell. By "steady-state dynamics" I refer to those self-regulating systems of interacting enzymes whose overall state is determined (and transmitted) by the relative concentrations and activation states of its constituents and is itself determined by phosphorylation state, glycosylation state, proteolytic processing, and the like. By "chromosome marking" is meant the pattern of modification of DNA through the addition of methyl groups to C (cytosine) bases, which affects the transcriptional activation state of DNA.

The treatment of these categories here departs significantly from that of Jablonka and Lamb in certain ways. Whereas their emphasis was on the role of chromosome marking as a kind of "shadow government" (my term) within the nucleus, the emphasis now will be more on the first two categories. With respect to organizational structure, I will offer, at least in miniature, a systematic discussion of the nature of membrane-based cellular compartmentalization with an emphasis on it negentropic properties and its fundamental and irreducible role in sustaining biological order. With respect to steady-state dynamics, this discussion will not be limited to its being only a *sub*-system; but will also focus on Stuart Kauffman's ambitious attempt to account for the emergence of biological order writ large through theoretically modeling complex-system dynamics. The purpose and goal of discussing these epigenetic inheritance systems will be to show that the relationship of each of them to the genome (and to each other) is that of codependence and causal reciprocity. To the extent that genes can be said to carry information at all, it will be argued, it is only in the context of those organizational structures and dynamics which are, in all respects, co-original with the genome and for which the genome itself can never substitute or prefigure.

Slippery Borders, Moving Membranes, and the Compartmental Division of Labor

Eurkaryotic life depends upon the spatial and temporal organization of cellular membrane systems.
—James E. Rothman and Felix T. Wieland, 1996

The gene-oriented rhetoric of life, which is broadcast and amplified by mass media, has rendered much of what masquerades as basic and clinical genetics into household vocabulary. And yet, the most basic concepts of cellular structure remain largely unknown to all but specialists. Lay people typically recall, usually from high school biology, something of an inventory of subcellular organelles, but they do so only as arbitrary facts lacking the systematic relationships which would allow one to make further inferences. Part of this discussion is intended to be basic and remedial; it seeks to provide the reader with at least a baseline understanding of cellular structure and its dynamics.

All living organisms are currently grouped into three categories: eukaryotes, eubacteria, and archaea. The latter two of these are strictly one-celled [although capable of social aggregation and complex pattern formation (Shapiro 1995)], lack a nucleus and other well-defined internal structures, and are thought to predate eukaryotes by 3 billion years. Eukaryotes, which include all multicellular organisms, as well as a diverse array of one-celled organisms (protozoans, yeast, and so forth), are the result of symbiotic "experiments" between eubacterians and archaeans. The eukaryotic cell has a diameter of about 10 times that of eubacteria or archaea and also has a complex internal organization. Part of this complex organization is the inclusion of bodies some of which (mitochondria, plastids, and possibly microtubule organizing centers) are the descendants of what were once free-living eubacterians.

Cellular identity, i.e., the demarcation between the inside and the outside of a cell, is constituted by an amphipathic boundary known as the "cell membrane" (or plasma membrane). Life as we know it is ever and always a water-based phenomenon. As in offset printing, life retains its boundaries on the basis of the immiscibility of water in oil. An "amphipathic" substance is one that has both water-miscible (hydrophilic) and water-immiscible (hydrophobic) components.

Cellular membranes are the aggregates of amphipathic molecules. The standard membrane-forming molecule is a phospholipid—a molecule with a hydrophilic phosphodiester head and a hydrophobic lipid tail. Cellular membranes then consist of sheet-like bilayers in which the center of the sheet is composed of lipid tails, whereas the surface of the sheet on both sides consists of "phospho" heads. The hydrophilic heads then intervene between the oily, hydrophobic core and the aqueous environments of both the internal milieu of the cell and the extracellular world (whatever that happens to be).

The force which affects the amphipathic boundary is just the same as that which causes water to ball up into droplets on the surface of a water-resistent cloth. Nonpolar, hydrophobic surfaces, like those of a water-resistent material, do not actively repel water, but neither do they provide the opportunity for stabilizing weak bonds.

Within the water molecule there is a separation of positive and negative charge. Surfaces in which there is also separation of charge offer the opportunity for transient but stabilizing weak attractive bonds with water. In the absence of electrostatic attractions, resistance to the loss of entropy holds sway. A layer of water one molecule thick, which is thereby all surface and has no interior volume, is considerably more organized than a spherical droplet of water, which minimizes its surfaces to volume ratio and maximizes the diffusional mobility of constituent molecules. Within the sphere water molecules can move fully in all three dimensions. On a flat surface diffusion is confined to two dimensions only and is thus far more organized and predictable. The force that results in water balling up on a hydrophobic surface, then, is the same as that which allows amphipathic membranes to constitute the principal boundary defining the "material" of the living world. Phospholipid membranes provide the principal partitions between cell and outer world but also serve to partition cells into distinctive intracellular aqueous compartments.

The fact of a plasma membrane that defines the boundary between the inside and the outside of a cell, while wholly indispensable, does not yet undermine Schrödinger's claim. A phospholipid bilayer is in itself, by any measure, information-poor. It is a fluid structure with little resistance to random motion in two dimensions and can be readily prepared in the laboratory. In order to make the case for how membrane-based cellular structures constitute a system that is independent of, and causally and functionally parallel to, as well as an equally basic source of biological order as, the genome, I will have to say more. I will provide evidence to show that (1) the membranous organization of the cell is that of a highly complex structure based on the differentiated inclusion of proteins and (2) that the movement of these proteins in the plane of the membrane is not random but itself a source of biological specificity. Further (3) not only is the orderliness of the membrane structure not dictated by "genetic information" but membrane and genome organization are complementary and mutually dependent, in effect, co-constitutive sources of cellular information.

The principal membranous bodies in a eukaryotic cell can be placed into two categories. First, there are those fully enclosed spherical-to-oval bodies: the mitochondria (in all cells) and plastids (in photosynthetic cells), which are the remnants of once free-living prokaryotes. Second, there is the interdependent network of irregularly shaped membranes extending from the nucleus to the plasma membrane in an essentially concentric fashion, including a variety of associated vesicles. Subsequent discussion will focus on the latter system only.

The principal membranous system of the cell—and it is very much a system—consists of something like a series of pancakes bent around the center of the cell and extending one after the other toward the cell surface. Whether these pancakes form something like concentric rings (albeit with occasional bites taken out of them) depends on the developmental state and tissue type of the cell. A cell which has assumed a certain polarized morphology may have its nucleus down on the basal side of the cell, with a complete nuclear membrane pancake surrounding it but with subsequent "stacks" extending only away from its basal side rather than radially in all directions. In any event, there is always a vectoral relationship between the membranous stacks, beginning with the nuclear envelope and ending at the plasma membrane. The nature of this vectoral relationship will be described here.

The membranous pancakes of the cell partition its contents into "luminal" and "cytosolic" domains. All the interiors of the pancakes are luminal. All that is not within the membranous pancake is either cytosol or nucleus. The nucleus is not within a membrane-bound pancake but rather is surrounded by the innermost layer of pancakes which wrap around it. The enclosure of the nucleus is not complete, however—its openings are referred to as "nuclear pores." The nuclear pores serve as gatekeepers in regulating which large molecules in the cytoplasm and nucleus can transit between these compartments. Water-soluble molecules, if sufficiently small, can pass freely through the nuclear pores, but the passage of larger molecules is regulated by chemically specific criteria. The inner face of the pancake that surrounds (and thus defines) the nucleus is referred to as the "nuclear envelope," as distinct from the outer surface, which is endoplasmic reticulum. Within the pancake is the water-soluble lumen of the endoplasmic reticulum. Between the endoplasmic reticulum and the plasma membrane of the cell is situated a complex stacklike series of pancakes referred to collectively as the Golgi complex. These are further differentiated into *cis, medial,* and *trans* golgi, with *cis* being close to the endoplasmic reticulum and *trans* farthest away toward the cell surface. Additional subcategorizations of the Golgi are variably distinguished by different cell biologists and for different cell types.

Proteins are universally recognized as the principal determinants of biological structure, function, and specificity. Proteins (1) provide the catalytic sites of almost all enzymatic reactions, (2) are the principal constituents of all forms of muscle, (3) are key to most immunological and other highly specific receptor-mediated recognition processes, and (4) provide the durability of hair, nails, and skin for protecting the surface of the body, the microskeletons within cells, the collagen matrix of connective tissue, and much else. It is *as* a store of template information for synthesizing proteins (and RNA) that DNA is accorded its function and importance. The achievement of the biological function of a protein is contingent not only upon its correct synthesis but also equally upon its post-translational modifications as well as its localization within the organism.

At the most general level of analysis proteins are located in four principal domains. (1) They are embedded within membranes, (2) they are resident in the lumen of various membranous pancakes, (3) they reside within the cytosolic or nuclear regions of the cell, or (4) they are excreted into extracellular domains of the organism. (This would include antibodies, clotting factors, enzymes, peptide hormones, etc., in the blood and lymph, the fibrillar matrices of connective tissue, and so on.) The system of cellular membranes, ranging from the nuclear envelope to the plasma membrane, is biochemically distinguished first by the composition of the proteins embedded in the membranes and secondarily by the composition of the proteins within their respective lumen. Maintenance of the differentiated identities of components of the cellular membrane system is requisite to the life of the cell and the life of an organism. Of the four categories of possible protein destinations mentioned above, three of them require transit through the membrane system.

Proteins which will remain resident in any membrane, which will remain resident in the lumen of any membranous body, or which will be excreted into the extracellular milieu all enter the membrane system at the same port of entry, the endoplasmic reticulum (ER). Membrane lipids too are synthesized at the ER, so the flow of biosynthetically new materials in the membrane system is vectorally directed outward, i.e., from the most central membranous body radially outward through each successive pancake in the stack and toward the plasma membrane.

Protein synthesis is initiated in the cytosol of the cell when a messenger RNA molecule has passed through the nuclear pore and has triggered the assembly of a ribosomal translational complex. Proteins destined for passage through the membrane system are equipped with a "signal sequence" at the N (for amino) terminal end of the polylpeptide chain, which is the first part to be synthesized. The appearance of the signal sequence halts further synthesis. For protein synthesis to resume the signal sequence must become associated with the surface of the endoplasmic reticulum, specifically with "docking proteins" embedded in the ER membrane. While the presence of a signal sequence may be said to be encoded in DNA, its function can be realized only in the context of a receptor complex already present in the ER. These receptors themselves would be "coded for" with signal sequences, and yet they must always be dependent on the presence of receptors already being embedded in the ER in order to receive them.

In a pattern that will be shown to be more elaborate, the differentiated structure of the membrane system constitutes the template for its own renewal. Genetically coded target information is only meaningful in the context of the already existing template of the differentiated membrane system which interprets it. Passage from the endoplasmic reticulum to the *cis* golgi and from one pancake in the stack to the next occurs by way of transient transport vesicles. Small vesicles with specific contents "bud-off" from the ends of the pancakes, only to fuse with the next pancake in the stack. In this way there is a steady flow of membrane directed outward toward the plasma membrane. Proteins to be secreted outside the cell are first deposited within the endoplasmic reticular lumen and then progressively transported by one vesicle after another until fusion with the plasma membrane excretes them into the extracellular environment. Proteins destined for inclusion in a certain membrane become embedded in the endoplasmic reticulum membrane; then, within the membranes are transported by way of each of the successive transport vesicles. Transport vesicles are each limited to communication between two levels, i.e., two pancakes. A class of recognition proteins referred to as SNARES (soluble NSF [*N*-ethylmaleimide-sensitive fusion protein] attachment protein receptor) largely mediates specificity of transport function. These are subdivided into the receptors which are present in the vesicle, referred to as v-SNARES, and those in the target pancake referrred to as t-SNARES (Rothman & Wieland 1996). As in the case of the docking proteins of the ER, t-SNARES must be present in the specific target membranes, marking the different "addresses" at each sequential level, for the continued self-renewing cellular assembly to proceed. The differentiated distribution of receptors in the membranous compartments of the cell are preserved through cell division, in perpetuity, for all succeeding members of the cellular lineage. (Only from cellular order comes cellular order.) Newly synthesized proteins embarking on their transit through the membranous system of the cell are endowed with endogenous sorting signals that most often consist of 4 to 25 residues but may also be determined by three-dimensional conformation (Rothman & Weiland 1996). Endogenous signals may either specify association with a v-SNARE and thus inclusion in a nascent transport vesicle and transit to the next locale, or they may serve to restrict inclusion, possibly by leading to association with membranous regions or patches that are refractory to inclusion in a transport vesicle. The conditions for retention are also likely to be differentiated features of the particular membrane location (pancake) which must be preserved and passed on across cell divisions and organismic generations. Proteins that lack any specific transit signal will simply travel by bulk flow according to its concentration in the donor compartment.

The movement of selected components from out of a membranous compartment (pancake) begins with the attachment of "coat proteins" to the outer (cytoplasmic) side of one region of the pancake. (This is how the budding off of a transport vescicle begins.) The coat proteins are recruited from the cytoplasm and consist of repeated units of the same protein that ultimately form a spherical shell around the emerging vesicle membrane. The v-SNARES that will direct the vesicle to its next compartment after it buds off the "donor" pancake become concentrated in that region surrounded by coat protein, as do other transiting proteins which contain the appropriate signals.

The mechanics of budding off are realized through the effects of the coat proteins whose polymerization on the membrane surface results in forcing it into a kind of droplet that can close onto itself as a sphere and become released. Assembly of cell surface-coat proteins requires the use of cellular energy stores by way of the degradation of high-energy guanosine triphosphate (GTP) molecules. Movement of the new vesicle to the next compartment generally depends on diffusion alone. (This is not the case for the transport of synaptic vesicles carrying neurotransmitters where microtubule tracks are deployed.) Specificity of contact is provided by the recognition reactions between the v-SNARES on the vesicle and the t-SNARES on the target pancake. The fusion of vesicle with target membrane is thermodynamically unfavorable. The energy barrier to spontaneous fusions protects the highly differentiated and information-rich membrane system of the cell from entropic heat decay within that temperature range which is compatible with the life of the organism. Fusion of the vesicle with its target compartment membrane requires the release of coat proteins from the vesicle, the achievement of close apposition of membranes mediated by the specific binding of SNARES, the formation of a new complex of proteins including NSF and soluble NSF attachment protein (SNAP), which together mediate the fusion process. Realization of the fusion event requires the expenditure of cellular energy stores, this time in the form of adenosine triphosphate (ATP), the principal high-energy intermediate that serves as the common coin of cellular energetics. After fusion with a new compartment the contents of the vesicle become disseminated. The new compartment will then undergo its own transport vesicle formation in which a different (albeit overlapping) set of proteins will become enclosed according to the specific biochemical identity of that compartment.

Schrödinger's reflections on the requirement of heritable order lacked any conception of how fluid structures such as membranes could resist entropic heat decay. That the dynamics and renewal of cellular membrane systems and other structural systems require a constant expenditure of cellular energy provides a strong indication of the extent to which these systems are principal sources of biological order and information unto themselves. Recognition of the self-templating, highly differentiated, decay-resisting, far-from-equilibrium nature of the membrane system, as well as other organizational structures, should be of no small significance in reconsidering Schrödinger's argument for why a hereditary code-script *had* to be the self-executing governor of all cellular-organismal processes.

The system of membranous bodies described as being biochemically distinct is also functionally differentiated. Each level in the stack has specific biosynthetic capacities. I have already argued that the biological "meaning" of a protein is not realized simply at the level of its amino acid sequence but is dependent on its localization in a particular cellular, or extracellular, compartment or milieu. In addition, the biochemical and cellular significance of a protein is highly affected by its post-translational modification, i.e., its acquisition of additional covalent bonds to carbohydrates, lipids, phosphate groups, and so on. Extending outward from the surface of all cells is a corona of mixed and variably charged oligosaccharide chains and proteoglycans, generally referred to as the "glycocalyx." The processes of multicellular development and differentiation—that is, organismic ontogeny—entails dynamic interactions between cells which result in the induction of new cell states (and sometimes cell death), in cell proliferation, as well as in the reproductive quiescence and stabilization of induced states during the formation of tissues.

The glycocalyx is critically involved in the "sociality" of cellular development. It also plays an integral role in the cellular interactions that induce changes in cellular states, and in turn it is subject to being chemically transformed by cells undergoing changes of state. Even terminally differentiated cells which are stabilized within some tissue matrix display cell-surface receptors that are responsive to blood-borne chemical messengers such as insulin, growth factors, and the like. Cell-surface receptors are themselves glycoconjugates, i.e., protein molecules with oligosaccharide attachments. The biochemistry of the oligosaccharides may significantly impact the binding properties of the active site of the receptor.

The construction of the glycocalyx occurs in a stepwise fashion. The elaboration of oligosaccharide attachments to proteins and lipids takes place as the proteins and lipids pass through the membrane system just described. Each compartment, each "pancake in the stack," has its own distinctive biosynthetic capacity. The glycoprotein chain is modified according to sequential exposure to the glycosyl-transferase enzymes of the respective compartment and the availability of saccharide primers.

The possibilities of oligosaccharide chain elongation are highly complex. Sugar units (the building blocks) may be removed as well as added along the way, chain branching may or may not occur, or different sugar-adding enzymes (glycosyltransferases) may or may not be present in a certain compartment at a certain time, and alternative glycosyltransferases may compete for the same growing chain. With all of these variables, subtle diffences in the cellular context may influence the glycosylation pathway. Changes in cell shape, for example, that would be influenced by cell-cell or cell-matrix attachments could affect glycoslyation patterns. But glycosylation patterns, which alter the glycocalyx of the cell, in turn can affect cell-cell and cell-matrix adhesion and thus the inductive state of the cell, the shape of the cell, and so on. One can envisage complex feedback loops that may become established with or without the involvement of changes in the transcriptional activation state of DNA. Further, one can plausibly postulate the possibility of "organizational mutations" which become stabilized and propagated during the subsequent developmental history of the organism and even conceivably transmitted to progeny.

Thus far I've described the differentiation along the *radial* axis of the principal system of cellular membranes. Now we turn to a consideration of the organization and movement of proteins in the plane of the membrane (*orthogonal* to the axis just discussed); in so doing we will draw on some theoretical work of Max Delbrück. Whether Delbrück conceived of this work as continuous with, and a logical extension of, the issues broached earlier by him and Schrödinger I can't say, but I hope that it will become evident why the suggestion of such continuity is warranted. The membrane dynamics to be discussed focus on experimental work performed on the plasma membrane, but there should be no reason why the findings couldn't pertain as well to the many internal pancake membranes that are not as readily accessible to experimental inquiry.

In 1972 Singer and Nicolson first described the concept of cellular membranes as bilayers of amphipathic phospholipids with embedded proteins as the "fluid-mosaic" model. It had already been shown by Frye and Edidin (1970) that if a mouse and human cell are fused (by use of a Sendai virus "fusogen"), both the mouse and human antigens become quickly mixed together and distributed generally around the surface of the "heterokaryon" cell produced. Cell-surface antigens are generally taken to be glycoproteins and glycolipids embedded in the membrane, and these early results were suggestive of free diffusion in the plane of the membrane. In 1975 Max Delbrück and a colleague provided a theoretical model for the diffusion of proteins, approximated as cylindrical objects diffusing freely in a membrane. For a cylinder with radius a diffusing in a viscous sheet with thickness h and viscosity μ, which is bordered by a less-viscous fluid of viscosity μ', the coefficient for lateral diffusion D_L is given as $D_L = K_B T/4\pi\mu h(\ln \mu h/\mu' a - \gamma)$ with K_B = Boltzmann's constant, T = temperature, and γ = Euler's constant (~0.5772) (Saffman & Delbrück, 1975). Using "ballpark" values for the respective parameters, a = 25Å, h = 35Å, μ = 1 – 10 poise, μ' = 10^{-2} poise, $K_B T$ = 4×10^{-14} ergs, results in a predicted D_L of 6.0×10^{-8} to 6.0×10^{-9} cm^2/sec (Cherry 1979).

I have previously indicated the strong evidence for the *nonrandom* distribution of proteins at the different levels of the membranous stacks that comprise the principal membrane system of the cell. The Saffman and Delbrück model provided a handle for determining whether the movement of proteins within the plane of a membrane is purely diffusional (and thus random) or constrained in ways that could prove to be of biological significance. One can see how Delbrück's exploration of this question follows the same lines of interest expressed by Schrödinger concerning the physical basis of the preservation of biological order. Random patterns of distribution are information-poor and require little by way of explanation. Highly ordered and specific patterns of organization, as we've seen with respect to the radial or concentric pattern of compartmentalization in the cell, do pose explanatory challenges with respect to the preservation of information. We will now consider evidence as to whether the movement of proteins in the plane of the plasma membrane—and thus largely orthogonal to the direction of the differentiation of the membranous strata—also poses such a challenge.

The principal technique that has been used for examining the lateral movement of membrane-embedded constituents has been fluorescence recovery after photobleaching (FRAP). In this method, cell-surface components are labeled with a fluorescent dye. A laser is focused on a small (1 to $10 \mu m^2$) area on the surface of the labeled cell. Fluorescence in this area is monitored by a photomultiplier. By momentarily increasing the intensity of the laser by 10^3- to 10^5-fold the fluorophores in the region can be photochemically bleached. The laser is then attenuated and fluorescence once again measured. The time it takes for full recovery of fluorescence (due to the diffusional replacement of the bleached fluorophores) is used to calculate the D_L of the membrane-embedded cell surface components. The level of fluorescence immediately after bleaching is designated f_0. The complete or highest fluorescent level is designated $f_{\infty.\forall}$. From the times it takes to recover maximum fluorescent recovery is derived the $t_{1/2}$, or half-recovery time. The diffusion coefficient, D_L, is then given by the equation $D_L = (\omega^2/4t_{1/2})\ \gamma$, where v is the radius of the beam, $t_{1/2}$ the half-recovery time, and γ a parameter which accounts for the degree of bleaching and beam profile. Under typical conditions $\gamma = 1.3$ (Cherry 1979).

Simple-model membranes can be constructed in the laboratory in which nothing is present above or below the plane of the membrane. Various proteins can be incorporated into these model membranes and the lipid composition can be altered in order to evaluate the influence of lipid composition on diffusional coefficients. The model membranes provided a good opportunity to test the predictions of the Saffman-Delbrück equation. In a number of FRAP studies using different lipid compositions and different proteins, D_L's were found to be in the 10^{-7} to 10^{-8} range, which is in fairly good accordance with the Saffman-Delbrück predictions (Smith et al. 1979, Vaz et al. 1981). The model membranes were also used to assess the effects on diffusion of the presence of hydrophilic components in the diffusing protein. Structurally, the membrane can be imagined as being like a thick sandwich with fairly thin slices of bread. The bread would constitute the hydrophilic parts of the membrane with the much thicker hydrophobic lipids sandwiched between. The hydrophilic aspects of a membrane protein (which may include oligosaccharide chains) will be "dangling" above or below the plane of the membrane, being significantly longer than the thin width of the hydrophilic part of the membrane. The model membranes allow for the effects on diffusion of the hydrophilic protrusions to be uncoupled from any effects due to the interaction between such protrusions above or below the plane of the membrane with other intracellular or extracellular components of a real cell. Comparison of the diffusion coefficients of the membrane protein gramicidin, which possesses no hydrophilic portion, with that of glycophorin, which has a large hydrophilic portion, revealed little difference, suggesting that hydrophilic protrusions from the membrane as such are not important for determining diffusion rates. This will be seen to become important as the possibility of diffusional constraints based on biologically specific interactions between hydrophilic groups and cellular or extracellular constituents outside of the membrane is considered.

FRAP studies carried out on a number of different living mammalian cells (in culture) revealed two major findings: (1) cell-surface proteins appeared to be present in a mixture of mobile and immobile fractions and (2) the mobile fraction is generally at least two orders of magnitude (100-fold) slower ($\sim 10^{-10}$ cm^2/sec) than the range of diffusion mobilities found in the model membranes. Subsequent studies provide strong evidence for the cause of both of the above to be based on constraining interactions between the membrane-embedded proteins and cellular constituents within and adjacent to the cytoplasmic face of the membrane. The possibility that a reduction of diffusion rate might be due to increased viscosity within membranes in which the protein-to-lipid ratio is very high was discounted by the finding that the photoreceptor pigment protein rhodopsin, which is packed at a maximally dense ratio of approximately one to one with membrane lipids in the outer segment of retinal rod cells, actually diffused at the comparatively rapid rate of $D_L = {\sim}3.5 \times 10^{-9}\,cm^2/sec$ (Poo & Cone 1974).

The advent of electron microscopy in cell biology revealed, among many other things, the presence of various kinds of filamentous structures within the cell, collectively referred to as "cytoskeleton." The presence of something like a cytoskeleton in cells—especially if it is highly specific in form with respect to cell type, developmental stage, and so forth—is clearly relevant to our concerns about structural information, its transmission, and its relationship to the genome. I will introduce the cytoskeleton now within the context of its putative role in constraining the free diffusion of membrane-embedded proteins.

Support for the cytoskeletal interaction with integral membrane proteins was provided by studies of diffusion in membrane "blebs." Blebs are protrusions of the plasma membrane that are thought to be detached from the underlying cytoskeletal connections. Blebbing is induced by various factors, such as cross-linking of membrane proteins, anoxia, physical injury, and prolonged protease treatment of a cell. Wu et al. (1981) induced blebbing in mouse lymphocytes and compared the difference of diffusion rates of both protein and lipids between the blebbed and normal cell-surface membrane. The diffusion rate of protein in the blebs was 1000-fold faster than in unblebbed normal membrane, where blebbing only enhanced lipid diffusion by a factor of 4. This difference lends support to the idea that membrane proteins, but not lipids, become associated with cytoskeletal structures which limit their mobility. What is not immediately evident however, is whether and to what extent the membrane protein-cytoskeletal interactions are biologically specific and information-rich in nature. This question was more directly addressed by studies that examined the partitioning of membrane proteins into mobile and immobile fractions.

Studies carried out on the NIH 3T3 mouse fibroblast cell line examined the mobility of two cell-surface receptors, the insulin receptor and receptor for epidermal growth factor (EGF). In both cases the mobile fraction was between 40 and 80 percent of the total receptor population (Schlessinger 1978). Interestingly, the mobile fraction plummeted toward zero when the temperature was raised from 23°C to 37°C. This was attributed to the likelihood of receptors being aggregated or internalized at the elevated temperature. The ability of a chemical messenger, such as insulin or a growth factor, to elicit biological responses in receptive cells is contingent on a cascade of events occurring after the chemical messenger is bound at the cell surface. These often involve the internalization of the chemical messenger–cell surface receptor complex. Specific internalization of bound receptors implies the presence of a cellular apparatus capable of selectively picking out the right molecules for internalization. Such mechanisms, which are the stock-in-trade of "signal transduction" processes, have long since been well characterized. We will consider further evidence for the specificity of membrane protein immobilization but first briefly comment on the possibly larger implications of the cases referred to above. It was reported that for both insulin and EGF receptors that raising the temperature from 23°C to 37°C resulted in a complete loss of diffusional mobility. What is of particular interest about this is that based on thermodynamics the loss of diffusional freedom with an increase in temperature is the exact opposite of what one would expect. In order for heightened temperature to result in the *loss* of entropy, the cell would have to push the process up a thermodynamic hill by expending its own energy stores.

The idea that an organism can buffer itself against noxious perturbations such as heat fluxes is of course nothing new. But how far down the organizational hierarchy would one expect this capacity to be found? The immobilization of cell-surface-membrane proteins in response to heat is suggestive of a biological stress response resulting in the loss, rather than the gain, of entropy. When considered down to the level of molecular dynamics within the cell, it suggests that *biological* resistance to thermodynamically driven entropic heat decay obtains all the way down to the most basic fabric of living matter.

Schrödinger had no window on the negentropic dynamics of *intracellular* molecular processes. His hereditary code-script vision is one which partitions the source of entropy resistance to within the nucleus, and the fact of such a partition continues to be implied by the rhetorical tradition that distinguishes the genome as *the* source of biological information. Empirically, I suggest, there just is no such partition to be found or any asymmetrical flow of "order." Schrödinger's order-from-order descriptor well characterizes the cell as a whole—but only as a whole.

The cell's system of membrane-based compartmentalization, post-translational modification, and transport provides perhaps the best sub-cellular analogue to Schrödinger's clockwork mechanism that operates *as if* it were invulnerable to thermal fluctuations. Like the clockwork the membrane flow system of the cell is in constant motion. Unlike the clockwork it resists heat decay not through the rigidity of its parts and steric constraints on their motions but rather through a colossal system of gated checkpoints. The cell in effect "plays off" of thermodynamic barriers. It uses high-energy thresholds, such as that of membrane fusion, to limit the occurrence of fusion events to those in which cellular energy stores are mobilized but under strict constraints delineated by the order which is heritably embedded in the system itself. The goal of this chapter will continue to be that of providing an overview of how biological order is multifaceted, distributed, and systematic.

Subsequent articulation of the premise that immobilization of membrane proteins is associated with the biologically specific binding of proteins to cytoskeletal or other constituents on the cytoplasmic surface focused on the use of the red blood cell for study. Mammalian red blood cells have been a mainstay of plasma membrane study because they are easily obtained and the plasma membrane can be readily isolated. Red blood cells, which are already enucleated (devoid of a nucleus), can be purged of their hemoglobin (their major cytoplasmic constituent) and yet will continue to retain the approximate size and shape of the original cell. Such experimental preparations have been referred to as red-cell "ghosts." The red-cell ghosts can be prepared in either normal or inverted, inside-out forms.

Red blood cells are required during the course of their journey through the circulatory system to undergo major deformations. A biconcave disk of 8 μm, it is drawn through capillaries of less than 2 μm in diameter (Goodman et al. 1983). The red cell's elasticity and resistance to mechanical damage is derived from a meshwork of protein filaments that adhere closely to the cytoplasmic side of the plasma membrane. The major protein of the meshwork, comprising 75 percent of cytoplasmic protein, is the heterodimer[1] "spectrin." While red cells have certain unique features and requirements, subsequent work has continued to find a wide variety of cells with elements in common with the red-cell membrane.

Differences between the membrane structures of various cell types occur along a continuum.

The predominant transmembrane protein of red cells is an anion channel referred to as "Band 3" and found to be present in approximately 10^6 copies per cell. Band 3 contains a large region that extends into the cytoplasm. The principal components of the "membrane skeleton" found on the cytoplasmic face of red cells include the α & β spectrin subunits, actin—which is a ubiquitous component of cytoskeletal structures (as well as muscle tissue)—and a protein called ankyrin. Comparison of spectrin binding to inside-out versus right-side out "ghosts" demonstrated the 10-fold greater preference for binding to the cytoplasmic surface (accessible on the inside-out preparation). Treatment of the cytoplasmic surface with a protein-degrading enzyme (chymotrypsin) resulted in the loss of 90 percent of the spectrin binding sites (Bennett 1978). The protein fragments released by chymotrypsin digestion were assayed for their ability to inhibit, through competition, the binding of spectrin to the cytoplasmic surface of the red-cell membranes. The fragment found to accomplish this was one derived from ankyrin (Luna 1979). Having established the principal linkage between spectrin and another (peripheral) component of the cytoplasmic membrane skeleton, similar methods were undertaken to establish that ankyrin is also the principal linkage between the transmembrane protein Band 3 and the cytoplasmic membrane skeleton (Hargreaves et al. 1980).

The image that emerges from these findings is one in which spectrin heterodimers form an intricate latticework on the cytoplasmic surface of the membrane with periodic attachment points to ankryin. Ankyrin serves as a kind of adapter molecule—itself being fastened to the membrane by linkages with Band 3—that traverses the membrane. The distinction between the mobile and immobile fraction of the Band 3 molecules could then be addressed by consideration of the stoichiometry or numerical ratios of the Band 3 and ankryin molecules. Band 3, as suggested above, was found to be present in approximately 10^6 copies per cell whereas ankyrin was found to be present in only about one-tenth of this amount. Band 3, however, is thought to form dimers and higher-order aggregates. The binding of Band 3 in the form of tetramers (groups of four) to single molecules of ankyrin would, for example, be consistent with the partitioning of Band 3 into mobile and immobile fractions along the numerical lines observed. Further support for this model was derived from subsequent studies on the nature of diffusional constraints on Band 3.

Without providing technical details, it was found that the rate of diffusion of the mobile Band 3 fraction and the ratio of mobile to immobile Band 3 fractions (i.e., the respective size of the fractions) could be uncoupled and modulated independently. While the rate of diffusion was shown to be a function of the steric hindrance of untethered Band 3 molecules by the spectrin lattice, the immobilization of Band 3 was a function of its being tethered by ankryin. This picture gained further credence through consideration of *rotational*, as opposed to *lateral*, diffusion. While the binding of Band 3 directly to ankyrin would inhibit rotational diffusion, the steric hindrance of the mobile Band 3 fraction would not be expected to affect its rotational movement. Nigg and Cherry (1980) examined the affects of proteolytically cleaving the cytoplasmic portion of Band 3 on its rotational diffusion and found enhancement of only 40 percent of the Band 3 molecules, a result consistent with the idea that only those Band 3 molecules (presumably about 40%) bound to ankyrin were rotationally inhibited. Since the time of these early pioneering studies on the structure of the red-cell membrane subsequent studies have demonstrated that the movement of proteins within the plane of the membranes of *all* cell types are regulated by highly specific, biologically significant mechanisms.

The standard rationale for speaking of genes in the conflationary style—as the "information," "blueprint," "program," "instructions," and so forth for building an organism—is that DNA provides the template for synthesizing proteins and that proteins, as enzymes, regulate all of the chemical reactions of the cell. For this rationale to hold up it must be the case that either (1) spatial arrangements of enzymes in the cell are of no great consequence or (2) that spatial arrangement is somehow prefigured and predetermined by the one-dimensional array of nucleic acids in the genes. A principal objective of this chapter is to provide evidence and an argument to the effect that neither of these is the case. This current rationale stands in a close and not accidental proximity to Schrödinger's argument. If Schrödinger was correct in assuming that thermodynamics prohibited anything outside of the aperiodic crystal from playing a central role in the continuity of living order, then the spatial arrangement of proteins in the cell could not in itself be of much consequence. Whether the movement from Schrödinger's thermodynamic argument to this standard rationale has ever been made explicitly or not, it is at least implicit in the continuity of conflationary gene-centered talk from the hereditary code-script through present-day programs and blueprints. Under the heading of "organizational structure" I have begun to marshal evidence on behalf of the idea that *cellular context* as a whole is basic to the nature and continuity of living beings and is irreducible to any of its constituent parts. The membranous system of the cell, the backbone of cellular compartmentalization, is the necessary presupposition of its own renewal and replication. Cellular organization in general and membrane-mediated compartmentalization in particular are constitutitive of the biological "meaning" of any newly synthesized protein (and thus gene), which is either properly targeted within the context of cellular compartmentalization or quickly condemned to rapid destruction (or cellular "mischief"). At the level of the empirical materiality of real cells, genes "show up" as indeterminate resources, that is, as kinds of Gene-D.

While even an uncontroversial depiction of the complexity and longevity of cellular structural organization is in itself enough to defeat Schrödinger's argument for why the aperiodic crystal *must* be *the* self-executing code-script, it is not yet enough to undermine a more discrete attribution of informational primacy to the genome. The structural organization of the cell, the basic membrane system, and the compartmentalization which it embodies is passed on from one generation to the next by way of the maternal egg cell. If cellular membrane organization is ever lost, neither "all the king's horses and all the king's men" *nor* any amount of DNA could put it back together again. But if the nature of cellular structural information is basically the same throughout the living world and cannot be used to distinguish between an amoeba and a human, then something like a modified story about genetic code-scripts dictating life-forms may still be defensible. On the other hand, if organismal genomes consist of a compilation of sequence motifs and exons, which are common throughout the living world with no species-specific stamp on them, then the onus of explaining where evolutionary innovation is to be found weighs even more heavily on the "gene-speakers". If indeed genes are basically interchangeable across kingdoms and phyla, as a surfeit of empirical findings attests to, then surely the specificity of organisms must be determined at a higher level of organization. If we are leveling our gaze at only that which is materially transmitted from one generation to the next through the one-celled bottleneck of sexual reproduction, then higher-level organization begins with chromosomal structure and ascends *only* to the level of the largely membrane-mediated topography and compartmentalization of the cell.

The potential for heritable structural alterations of cellular organization to have evolutionary significance is most accessible to investigation in the case of single-celled organisms in which there is no distinction between somatic and germ-line cells. Classic studies of this sort have been carried out on ciliates such as *Paramecium* and *Tetrahymena* (Jablonka & Lamb 1995). These cellular organisms are covered with rows of cilia, each of which is associated with a basal body and possesses a certain orientation. These units are asymmetrical and are the templates for their own replication. When the orientation pattern of the cilia–basal body is experimentally altered by environmental manipulation or microsurgery, the new pattern is transmitted to progeny. This pattern is preserved through both repeated generations of asexual reproduction as well as through sexual conjugation. Larger-scale patterns of heritable variation have also been witnessed, as in the case of the formation of "doublet" cells in *Paramecium tetraurelia,* experimentally induced by interfering with cell division. The doublet phenotype is then transmitted to subsequent progeny. Evidence for the natural emergence of a new species on the basis of structural mutation is cited by Frankel (1983) with regard to the ciliate *Teutophrys trisulca,* which possesses a single trunk but three anterior prosboces, each of which is similar to the single proboscis of the related species *Deleptus. Teutophrys* came about, it appears, through a structural mutation that produced the triplet organization and that was then perpetuated by the epigenetic inheritance of structural organization and eventually stabilized by genetic changes.

The basic character of metazoan development speaks to the plausibility of the idea that structural changes in the cell can be of evolutionary significance. Metazoan ontogeny consists of the progressive differentiation of cell lineages which, once differentiated, reproduce true to the identity of the lineage. If one were to look at ontogeny as a model of phylogeny (not exactly novel), one would see the same exact genes situated in different cellular contexts of different cell lineages, giving rise only to progeny determined by the cellular contexts. Once the membrane system and cellular organization of a cell are differentiated along certain lines, they become a stable basis for maintaining themselves and reproducing subsequent generations of the same cell type.

Helen Blau at Stanford carried out numerous experimental examples of the ability of the cellular context to condition the differentiation state of even foreign nuclei (Blau et al. 1983, Blau et al. 1985, Miller et al. 1988). Blau's experiments consisted of using a "syncytial" muscle fiber that comes about naturally through the fusion of muscle cells and is thereby multinucleate. Nuclei from other tissues types of the same or other species can be experimentally transferred to a cultured muscle fiber in order to see how the muscle cytoplasmic "context" will affect, for example, the transcriptional activity of a nucleus derived from a liver cell. Using liver nuclei from a different species enables the investigator to readily distinguish, by immunological means, which newly synthesized proteins in the muscle fiber were derived from the genetic templates of a muscle nucleus from the "host" cell versus those of the "donated" liver nucleus. When a nucleus from a liver cell of species A was transferred into the muscle syncytium of species B, Blau found that cultured syncytium produces muscle proteins, but not liver proteins, with the immunological identity of species A. The ability of a somatically heritable cytoplasmic context to regulate genomic expression, as demonstrated by Blau, certainly further suggests the possibility that epigenetic changes which are heritable across generations may be the source of evolutionary innovations.

Of course the real question concerning metazoan ontogeny is just how a single cell gives rise to the requisite number of differentiated cell lineages with all the right inductive developmental interactions required to reproduce the form of the mature organism. Understanding the dynamics between different components of the fertilized egg cell (and its surround) that become the developmental pathway of the nascent organism is still a central challenge. It is well established that the compartmental-ization of the cell in general and of its messenger RNA—the legacy of the maternal egg cell—is extremely influential in setting early develop-ment along a certain course. That this organization is not merely the product of nuclear inheritance but also significantly of structural inher-itance is directly addressed by Grimes and Aufderheide (1991) in their extensive review of the subject:

> The highly organized cytoplasm of a metazoan egg, therefore, *cannot* be solely the consequence of direct nuclear gene activity. Given the background of infor-mation from the Cliophora, one would predict that structurally heritable infor-mation systems must be present in addition to direct nuclear (genic) control systems in the metazoa. The ciliated protozoa are a group of organisms that have made exceptional use of the posttranslational, 'epigenetic' systems that con-tribute to the localization of gene products . . . Processes homologous, or at least analogous, to the directed assembly and directed patterning seen in ciliates are also functional in metazoa, and are of fundamental developmental significance (p. 67).

Order from the Inheritance of Self-Sustaining Dynamics and/or the Emergence of Self-Sustaining Order for "Free"

The Global View

In his contribution to a fiftieth anniversary retrospective of Schrödinger's *What is Life?*, Stuart Kauffman (1995) points out that the force of Schrödinger's argument was based on the assumptions of equilibrium thermodynamics held by "most physicists of his day." Macroscopic order, in this view, is attributable to " averages over enormous ensem-bles of atoms or molecules . . . not to the behavior of individual mole-cules." At equilibrium the predictability of the location of the components in a system in relation to one another is very weak except for the case of those atoms that are held together in a crystalline array. If given the presuppositions of equilibrium statistical physics and asked to account for the stability of complex life-forms through and across gen-erations, then it would follow that some form of solid-state structure, i.e., a crystal, and an aperiodic one to make it more interesting, must be involved. But life is not an equilibrium phenomenon, and so-called dissipative far-from-equilibrium systems are not only capable of sustaining complex highly ordered structures and dynamics outside of the solid state but also of self-organizing into other and even more highly ordered organizational regimes.[2]

The idea that stable biological order can emerge from the dynamic interactions of catalytic molecules goes back to the advocates of the "protein-first" model for the origins of life in the 1920s (Moss 1999). Following that tradition, Kauffman (1993, 1995) has argued that an aperiodic crystal, that is, the securing of genetic information in some solid-state array, is neither necessary nor, for that matter, sufficient for the existence of a stable system capable of evolving by the differential selection of heritable variations. Life emerges, in his view, on this basis of a phase transition in which a hitherto chaotic milieu of molecules in a thermodynamically open system self-organizes into an autocatalytic cycle. An "autocatalytic cycle" is one whose components cyclically catalyze their own production with their respective concentrations being maintained at a dynamic equilibrium over time. Where origins-of-life theorists had previously conceived of the emergence of life as a very low probability event, Kauffman (and collaborators) have used computer simulations to show that the state-space of a system of reactors and reactants will, given certain parameters, predictably converge on an autocatalytic cycle through a limited number of states which constitute an attractor cycle for the system. Kauffman thus shifts the locus of biological order and stability from that of solid-state bonds to the dynamics of systems whose state-space converges on attractors. That the latter can occur without anything like a genetic code is given plausibility by the simulations which Kauffman and Sante Fe Institute collaborators have produced. An independent group of investigators in Japan have likewise used computer simulations of simple reaction-diffusion systems to show that patterns resembling the heritable differentiation of cell types can be the result of dynamic phase transitions in the absence of anything playing the role of genes. In addition, Kaneko and Yomo (1994) were able to correlate trends in their simulations with certain biological phenomena observed in bacterial cultures.

What then is the relationship between DNA—i.e., the aperiodic crystal—and steady-state dynamics? Kauffman emphasizes that the former is neither necessary nor sufficient for the latter—which is to say that homeostatic, autocatalytic systems with the capacity to mutate and become subject to natural selection are at least theoretically independent of the need for any noncatalytic, chemically inert template polymer (e.g., DNA). Further, the presence of such a polymer cannot and does not specify (or guarantee) any particular dynamic regime. The presence of DNA or any other aperiodic crystal, for example, cannot itself prohibit a slide into chaos. DNA can at best be considered a kind of "fellow traveler" in the ship of life.

Mutation, defined in the dynamic-systems context, refers to a fluctuation in the synthetic catalytic cycles resulting in the accumulation of a novel product. Such a product may then become stably incorporated into the cycles with an alteration in the catalytic dynamics. Such dynamic mutations were evidenced in simulation. Now clearly, biological entities, as we know them, consist of DNA templates (Genes-D) *in the context* of steady-state dynamics so that attempts to distinguish between them, beyond just being arbitrary, run the danger of becoming a category mistake. Genes(-D), as real biological effectors, are the result of dynamic processing on multiple levels (e.g., transcriptional regulation, transcript processing, transcript transport, and translational regulation, etc.), so the very concept of the gene brings with it dynamic presuppositions. And yet the challenge of how to simultaneously cognize sequence-based and dynamic aspects together persists. Just as in the case of our distinguishing heritable *structural* features of the cell, the sense of referring to steady-state dynamics as an "epigenetic inheritance system" can properly be meant only to *analytically* distinguish concurrent and mutually dependent aspects of integrated living systems.

Kauffman attempts to elucidate this relationship through a theoretical simplification that appears to reap certain rewards. Following the considerations touched on above, pertaining to the number of types of molecules and the number of types of reactions, Kauffman centers on the ramifications of these two parameters for self-organization. In the first approximation, N represents the number of components of a living system and K the number of connections between them. The trick for Kauffman will be to parse the cell in such a way as to result in values for N and K that have interesting implications. To ask how many components there are in a cell is akin to Wittgenstein asking how many things there are in a room. It entirely depends on how you divvy up the room—and thus decide what is going to count as a "thing." And then depending on what is thus counted as N will follow how many interactions there are between the elements of N, i.e., K. Now if each component has only two states—i.e., can be counted as a binary variable, but receives inputs from a $K = N$ number of factors—then even where $N = 200$ the number of possible states of the system would be 2^{200} which is so sufficiently large that even if the state changed only every 1 to 10 minutes, it would still require more than the age of the universe for the system to sample every state.

Such an attractor cycle could not undergo selection, for obvious reasons. Kauffman elects to limit his world of N-relevant components to genes, settling upon 100,000 as a plausible number. As we've said, if $K = N = 100,000$, i.e., every gene affects every other gene, then the number of possible states would be far off of any relevant scale, and the succession of states would be random and chaotic. Having selected genes as the basis of his N parameter, Kauffman can entertain the abstraction of an organism as a Boolean network with 100,000 binary variables. Order emerges from such a system if the value for K is 2 or slightly less. With $K = 2$ the system is highly interconnected and complex, such that the state of variables is far from independent of one another, yet it is simple enough for discrete patterns to emerge as opposed to being condemned to a wholly unwieldy chaos. In the Boolean model, the activation state of each gene can be computed on the basis of one of 16 possible Boolean functions (randomly assigned)—e.g., AND, OR, IF, and so forth—as well as input received from ($K =$) two sources. Now given such an approximation, Kauffman finds the very satisfying result (through simulations) that the number of states which make up the state cycle approaches the square root of the number of variables. So, with 100,000 variables the predicted number of states in a cycle would be on the order of 317, which is roughly the number of differentiated cell types that cell biologists have distinguished in humans. The convergence of the state-space of this hypothetical system with 100,000 variables down to 317 possibilities is what Kauffman refers to as "getting order for free."

The pertinent question, however, will be if and to what extent the parameters that Kauffman has chosen can be understood to be of real biological relevance.

Kauffman's theoretical assimilation of the cell-organism to that of a parallel-processing genetic regulatory system has much of the character of a Faustian bargain. The idea that the biology of a living cell is more than just a series of linear reactions—more systematic, interconnected, and complicated than even a compilation of ever so many Rube Goldberg schematics—is not so much controversial among biologists as it is seemingly intractable. Kauffman's model provides the rare handle for conceptualizing biological processes at a higher level of complexity. It brings with it a power evidenced in Kauffman's ability to serve up explanatory accounts of phenomena ranging from the origins of life to ontogenetic differentiation and morphogenesis (Kauffman 1993). But at what cost? In modeling the cell-organism as a complex genetic regulatory system, Kauffman, somewhat ironically, contributes to the attempts to make the real (extragenomic) complexity of life *disappear*. Now surely Kauffman's model is militantly anti-genetic-reductionist in the sense that its basic unit is not a gene but rather a cell-state defined in terms of which genes are turned on and which are turned off. While the activation state of a cell's genome is considerably less reductionist than merely a focus on individual genes, the identification of a cell's phenotype with the state of its genome—and by extension the identification of the phenotype of an organism with the genetic regulatory state of all of its cells—is insidiously seductive and patently false. While the activation state of a cell's genes is inseparable from its phenotype, it by no means uniquely determines a phenotype, for all the reasons discussed in section 2 above as well as many more. A pattern of gene expression, no less than individual gene expression, is only meaningful in the context of multilayered levels of organization, structure, and dynamics which are in no way reducible to patterns of gene expression. Kauffman, in his own way, is party to the kind of "slippage" discussed in chapter 2. For the sake of getting a grasp on a powerful theoretical engine, Kauffman has made the real biology of the cell, and thus *biology* itself, disappear into the virtual interstices lying between the connections of (genetic) nodes in a parallel-processing simulation.

The irony that I refer to above is as follows. Kauffman's slippage, his contribution to the disappearance of biology in the name of the genome, is the product of what is perhaps the most sophisticated and well-elaborated challenge to the "hegemony" of the genetic code-script. Against the heirs of Schrödinger, Kauffman denies the role of a solid-state set of instructions and offers instead a model whose order emerges, not out of the stability of covalent bonds at all but out of the higher order "logic" of complex dynamics at the edge of chaos. He has simply latched onto the genome for lack of any other theoretically amenable ledge to grasp onto amidst the biological maelstrom.

While the central theme of this chapter has been that of epigenetic inheritance systems—that is, stable sources of biological order which are inherited in *parallel* with the genomic sequence and codeterminitive of phenotype—Kauffman's understanding of steady-state dynamics, or really the convergence of the state space of a complex system onto a series of numerically, highly delimited attractor states, is *not* about a parallel system of inheritance. Dynamics is the big picture for Kauffman, and if we are to think of dynamics as epigenetic, then in the Kauffman picture, epigenetics always comes first. But perhaps a better characterization would be to say that genetics and epigenetics simply merge into one and the same. But a merging of the two is hardly unique in twentieth century biology, as chapter 1 revealed. Nor would we want to suggest that the distinction between genetics and epigenetics should be accorded the status of a natural kind (a natural distinction?).

At the end of the first chapter, I too suggested that ultimately a theory of epigenesis which allowed for a recontextualization of the gene *qua* Gene-D would subsume the legacy of Mendelian genetics and, thus, Gene-P. Has Kauffman provided this step into the future—subsuming the legacy of Mendelism, not into a more "reduced theory" but rather into a more holistic and expansive theory? Certainly not. Just as the Mendelian tradition found practical utility in proceeding as if genes directly determined phenotype, Kauffman has found theoretical utility in proceeding as if the cell were constituted by 100,000 binary units and as if each of these received input from only two other components. Kauffman's model is itself a form of instrumental reductionism. Perhaps we should consider it a brand of "instrumental dynamicism" as opposed to the instrumental preformationism of the Mendelian tradition. In both cases a wealth of biological complexity is made to disappear in order to provide a provisionally useful simplification. To what extent Kauffman's instrumental reductionism provides practical as well as theoretical utility is of course yet to be proven.

If we begin to look at what must be the enabling background conditions for mammalian organisms to be modeled as if they were $N = 100,000$ and $K = 2$ systems, we will find our way back to the *wet biology* of heritable steady-state epigenetic systems. Consider the meaning of $K = 2$. In what sense can it be said that a gene only receives input from approximately two other genes? Strictly speaking, genes do not receive input from any other genes without the mediation of proteins. And even if the intent is just to speak of those genes which give rise to the directly mediating proteins, then there is still a retinue of proteins and thus genes which mediate the production of these mediating gene products.

It is clear then from the outset that any talk about a highly delimited number of gene inputs must distinguish between a great number of gene-input candidates and find some criteria by which most of these may be bracketed and treated as if they were only background conditions. As discussed in the second section above, the meaningfulness of any gene depends on the complex organizational structures and compartmentalization of the cell, which determines where the protein will be located and how it will be covalently and noncovalently modified. It is only at this level of finishing that a protein takes on biological significance. Clearly, there are many untold genes associated with the biological realization of most any protein at this level, and so these all must be factored *out* of what would have to count for Kauffman as "genetic inputs." The obligatory move to make then would have to be to decree that *all* those genes associated with gene products that are required for the functional realization of any gene product would not count as a genetic input in Kauffman's sense. This would bracket out a great many (so-called housekeeping) genes. But would even this rather large concession bring Kauffman's requirements into the space of biological reality? Consider an exemplary case for the regulation of the activation or inactivation of a particular gene sequence by a small number of particular gene products and see what further heuristic concessions must be made to accommodate the Kauffman model.

Using recombinant techniques, Diamond et al. (1990) constructed a simplified version of the steroid-hormone regulation of the gene proliferin in order to analyze the additive and relational effects of three regulatory proteins that we may consider to be our Kauffman inputs. Glucocorticoid steroids influence genetic transcription when joined to a soluble glucocorticoid receptor and then subsequently bound as a complex to DNA enhancer regions known as "glucocorticoid response elements" (GREs). Transcriptional activation of a gene refers to the synthesis of an RNA transcript by an RNA polymerase enzyme complex, using the DNA "sense strand" as a template. The ability of the RNA polymerase to bind to the promoter region of the gene and begin synthesis depends on the configuration of proteins that are present and that may facilitate or inhibit transcription. "Enhancers" are regions of DNA that often involve context-sensitive (developmental stage, tissue type, and so forth) regulation of transcriptional activation. Diamond et al. (1990) simplified the in vivo biology of glucocorticoid regulation by inserting an attenuated GRE sequence of only 25 nucleic acid base pairs which, while capable of binding the hormone-receptor complex, reduces the number of other factors that may come into play. The transcriptional initiation factor AP-1, which is composed of two proteins, c-fos and c-jun, plays a mediating role in the transcription of many genes, but its role in proliferin transcription cannot be relegated to the background because it interacts in very specific ways with the GRE and receptor-hormone complex.

Using the simplified GRE construct in a model system, Diamond et al. explored the relationships of the glucocorticoid receptor-hormone complex and the c-jun and c-fos proteins in the regulation of proliferin transcriptional activation. We can view their results in terms of a $K = 3$ system by expressing their findings in terms of a series of Boolean operators with the additional simplification of transforming graded transcriptional effects into an all-or-nothing binary switch. The results would then be as follows. If complex and no (fos or jun), then negative. If jun and complex or no complex, then positive. If fos and jun and no complex, then positive. If fos and jun and complex, then negative. As an artificially simplified system, this model of transcriptional regulation should be viewed as being as simple as it gets, and yet even at first blush, it is more complicated than a Kauffman $K = 2$ system. In addition to the direct effects of c-fos, c-jun, and the glucocorticoid receptor, the model system also involves the glucocorticoid hormone that binds the receptor. The hormone is the product of many genes associated with the biosynthetic pathways of secretory cells in the adrenal cortex. Now, when the steroid hormone enters the target cell, i.e., a cell that has a receptor available, it binds to the receptor in the cytoplasmic compartment. Having done so, the receptor hormone complex may then gain entry into the nuclear compartment, which is as much as to say that only after binding the hormone can the receptor pass the entry requirements administered by the nuclear pore. And what about the regulation of the expression of the c-fos and c-jun proteins?

To trace even the more proximately antecedent determinants of these genetic inputs into the regulation of the proliferin gene is to embark quickly on an explanatory regress in which increasingly many contingent features of cellular organization—surface-receptor binding status, the phosphorylation states of cell-surface receptors and signaling intermediates, the presence or absence of a host of transcriptionally active effectors, as well as others—come into play. In the words of Keith Yamamoto, the leading investigator of glucocorticoid mediated regulation, the activation state of the proliferin gene is determined by "the complex state of the cell." To even begin to trace back the biology that must be presupposed in treating a cell as if it were a Kauffmanian genetic regulatory system is to rediscover quickly the wet biology of organizational structure and compartmentalization, real steady-state systems, and the close relationship between them. Whether Kauffman has ultimately provided a vehicle for conceptually grappling with the intricacies of real biology at a higher level of complexity (i.e., has given us the means to bring more biology into a single concept) or a more compelling pretext for simply ignoring the same, is perhaps the judgment around which the value of his work will turn. Maybe between these two extremes lies the possibility of conceiving of Kauffman's model, while far too bare (and dry-) boned to represent real biology, as a sophisticated metaphor, a symbol but not a substitute, for how local regimes of order emerge from ensembles of multitudinously differentiated, multitudinously interacting parts.

The Regional View

Kauffman offers a top-down approach to understanding the role of self-sustaining dynamics in the achievement, propagation, and evolution of biological order. Beginning with formalisms at a far remove from empirical particulars he attempts to incorporate the latter by reinterpreting them in the light of a global perspective. Differentiated cell types, for example, reappear as the allowed states of a $K = 2$ system. The alternative approach to conceptualizing the role of self-sustaining dynamics as a nongenetic–template-based source of biological order is one which attempts to build up from small-scale regional processes—and are thus not at all removed from empirical wet biology. This approach would be the one more in line with that of Jablonka and Lamb's (1995) notion of epigenetic inheritance systems (really, subsystems). There are many kinds of dynamic subsystems which would share the property of sustaining, even across generations, some physiochemical state based on the intrinsic ability of the system to adjust and adapt to contingencies.

Delbrück (1949) provided a small-scale theoretical model for a type of positive-feedback mechanism in which a metabolic pattern becomes heritable. In this model of his cyclically catalytic system there is some regulatory component (protein) the presence of which results in the promotion of its own synthesis (positive feedback). The metabolic state of such a system is thus highly sensitive to fluctuations in the concentration of that regulatory component such that the current state of the system would be an outgrowth of its past history. The "lac operon" of *Escherichia coli* (Novick & Weiner 1957) is an example of such a system. It was found that when cultured in low levels of lactose genetically identical *E. coli* bacteria diverge into two different heritably stable phenotypes. One phenotype synthesizes β-galactosidase and the other one doesn't. The key difference between these is based on chance fluctuations in the intracellular concentration of the permease enzyme necessary to transport lactose inducer into the cell. The permease gene is itself part of the lac operon. When a cell by chance expresses the permease in low-lactose concentration, it results in an increase of lactose uptake followed by the induction of more permease, then more lactose uptake, which leads to further induction until stable concentrations of permease and β-galactosidase are reached. (The latter enzyme breaks down the lactose.) New generations of *E. coli* in the presence of low concentrations of lactose will thus inherit either the β-galactosidase metabolism or the non-β-galactosidase–expressing metabolism. While this is a simple model, it does provide some insight into how the complex state of an organism can be the result of its dynamic (as opposed to conventionally genetic) adaptation to historical contingencies.

Self-maintaining (steady-state) regulatory loops need not necessarily entail alterations in transcriptional rate; they could also occur at the level of posttranscriptional regulatory processes. The so-called spliced state of an RNA molecule, for example, has been shown to be capable of self-sustaining, autoregulatory effects on subsequent RNA splicing (Bell et al. 1991). Proteolytic proteins—proteins which cut other proteins in specific ways—are themselves regulated by their proteolytic state. The activation of proteolyic protein, through its cleavage by another proteolytic enzyme, may initiate a cascade of proteolytic events that feed back on the proteolytic state of other tokens of that type. The inheritance of either RNA in a certain splice state or proteolytic enzymes in a certain proteolytic state will result in a daughter cell (including an egg cell) that is already poised toward a certain "dynamic trajectory."

Most ubiquitous, and yet theoretically untouched, is the role of phosphorylation states in mediating what appears to be *all of the decision points* in cellular life. Phosphorylation of proteins is accomplished by a class of proteins called "kinases" for which there are now thought to be 2000 related genes in higher organisms (Hunter 1995). The reverse reaction, i.e., the removal of the highly charged phosphate groups, is accomplished by a class of enzymes known as "phophatases." These are now thought to number in the 1000 range (Hunter 1995). Given the Human Genome Project's estimate of approximately 30,000 human genes (Lander et al. 2001), kinases and phosphates would account for nearly 10 percent of the entire human genome.[3] The activation state of both kinases and phosphatases is regulated by their own phosphorylation state.

The potential dynamic complexity of this subsystem alone is astronomical. The phosphorylation state of one or more phosphorylation-state effector proteins may be the key regulatory determinant of a complex autoregulatory system which is passed onto daughter cells. The responses of cells to all external and internal signals is mediated by cascades of phosphorylation and dephosphorylation reactions, which result in everything from determining anabolic versus catabolic metabolic patterns to "choosing" between paths of cell growth versus terminal differentiation. The vast complexity of the phosphorylation system must be a function of the stabilization needs of highly differentiated multicellular organisms. The challenge that lies ahead will be to determine how stabilization capacity is built into the dynamic architecture of the phosphorylation system itself.

A bottom-up approach to conceptualizing the steady-state dynamics of an organism would be one that manages to bring together the various subcellular systems into an integrated, dynamic whole. The idea of a cellular state as a basic level of biological integration and identity was already considered by Sewall Wright (1945):

Persistence may be based on interactions among constituents which make the cell in each of its states of differentiation a self-regulatory system as a whole, in a sense, a single gene, at a higher level of integration than the chromosomal genes. On this view the origin of a given differentiated state of the cell is to be sought in special local conditions that favor certain chains of gene-controlled reactions which cause the array of cytoplasmic constituents to pass the threshold from the previous stable state to the given one.

To what extent cellular states can be individuated from an empirical, bottom-up approach is an open question. Given the virtual infinity of potential cellular states based on theoretical combinatorics (Elsasser 1987) it is likely that the number of effective possible cellular states is in fact highly context-dependent (contra Kauffman). If some finite set of even context-dependent individuated cellular states could be identified and their properties of transformation and stabilization characterized, then it would be possible for the "stories" of both ontogeny and evolution to be retold, using dynamic cellular states as basic units of variation and selection. The course of an ontogeny would then be understood to be that of the reproduction of a long series of cell-state-inducing and cell-state-stabilizing interactions within multiple levels of constraining context. The ability of each set of state-stabilizing couplings between cells to become themselves part of the constraining context of subsequent cell-state decision points would have to be basic to the possibility of accounting for the macrostability necessary for producing whole lineages of generically similar organisms. The theme of cell-state variation and selection will be revisited from another perspective in chapter 4.

Chromatin Marking and the Fall of "Molecular Weismannism"

The rhetoric of the hereditary code-script as discussed in chapter 2 has helped itself to, among other things, the metaphorical resources associated with biblical religions. The "Book of Life" and other such textual tropes connote a deep sense of antiquity. A genetic code which is seen to have stood the test of eons is a likely candidate for accruing a secularized (albeit, barely) sense of sanctity. But such quasibiblical veneration trades on an ambiguity over form versus content of the code. Granted, all terrestrial life appears to be relatively united due to common descent as well as in terms of the form of the code, but it is only in terms of its form that the code can be said to be universal. The Ten Commandments, on the other hand, are esteemed for their antiquity and the presumed authority of their content, not for what language—whether Hebrew, Greek, Aramaic, Latin, or English—they are written in. DNA, for all its formal antiquity, turns out to be subject to rapid changes: transpositions, amplifications, recombinations, and the like, as well as modulation by direct chemical modification—all in ontogenetic as well as evolutionary time frames.

August Weismann held that the germ line of organisms is sequestered and insulated against the possible effects from the life experience of the host organism. Subsequent investigations of developmental patterns have since shown that the vast majority of organisms do not sequester their germ line early if ever in development (Buss 1987). Early sequestration of germ tissue appears only in higher organisms (and some invertebrates) and so cannot be treated as a basic evolutionary mechanism. In its place orthodox neo-Darwinian theory has substituted the idea that DNA is impervious to the effects of organismic experience; this has been referred to as "molecular Weismannism."

While it seems likely that a strong focus of twenty-first century biology will attend to the mechanisms of spontaneous recombination, the remainder of this section will consider the process known as "chromatin marking" (Jablonka & Lamb 1995, Jablonka 2001), whereby gene activity is modulated by direct chemical modification. Chromatin marking, also known as hypermethylation, is not in principle separate from the organizational and dynamic dimensions of life already discussed, but it does represent the most immediate epigenetic link between the historical, contextual life history of a cell-organism and the chemical structure of its genome.

Chromatin marking refers to the enzyme-mediated addition of methyl groups (CH_3) to the C (cytosine) bases in regions of DNA where C is followed by G (guanine), i.e., CpG dinucleotides. While (following Jablonka & Lamb 1995) I have grouped chromatin marking as one of three general classes of epigenetic inheritance systems, DNA methylation is generally what is meant in current biomedical literature when the terms "epigenetic programming" or "epigenetic mechanisms" are invoked. This is perhaps because the direct chemical modification of DNA would appear to correspond most closely to an etymologically literal interpretation of the word "epigenetic." (The semantic link between "epigenetic" and "epigenesis" is lost to most researchers and clinicians, who are woefully unschooled in the history of biology.)

The addition of the above methyl groups to the CpG dinucleotides within DNA decreases the likelihood of transcriptional activation, probably by inhibiting the association of DNA with proteins that promote transcriptional activity. Recent literature has also indicated an association between the methylation of DNA and the chemical modification of histone proteins which form the structural matrix of chromatin particles. The full significance of this with respect to epigenetic stability, heritability, and transcriptional repression is yet to be fully revealed. In addition to influencing transcriptional activation, the methylation state is also found to affect the susceptibility of DNA toward mutation, translocation, and meiotic recombination.

Chromatin marking results in context-dependent modulation of genome activity in two distinct ways. During the gametogenesis of mammalian eggs and sperm, chromosomes are methylated according to sex-specific patterns. The genes of the mammalian zygote are thus differentially predisposed to activation or inactivation depending on the parent of origin. This phenomenon has been referred to as "imprinting" and is understood to be responsible for the inability of mammals to reproduce parthenogenetically. Imprinting seems to ensure that the availability of both male- and female-derived chromosomes is necessary for successful development. Because the same allele (gene) at the same locus on differentially marked chromosomes can have different phenotypic consequences, the term "epialleles" has been introduced. Two epialleles can be associated with different phenotypes—not because of differences in their nucleic acid sequence but rather because of differences in their pattern of CpG methylation. For example, Prader-Willi syndrome and Angelman syndrome represent two phenotypically different human genetic diseases, which, as it turns out, are due to the same chromosome 15 deletion. In the case of the former it is associated with the paternal chromosome and in the case of the latter the maternal chromosome. Recent studies on Turner's syndrome individuals, females who have only one X chromosome (instead of two), have shown that it makes a difference whether the one X was derived from the mother or from the father. Those who derived their X chromosome paternally were reported to have a tendency toward better social skills acquisition. The phenotypic differences between Prader-Willi individuals and Angelman individuals and between different subclasses of patients with Turners are correlated with the possession of different epialleles. Huntington's disease, which is routinely touted as a paradigm example of the utility of classical genetics owing to its high rate of "penetrance," is in fact notably non-Mendelian in a number of ways, including epigenetic-based parent of origin effects.

The other pattern of chromosome marking is that which occurs, not during gametogenesis, but rather during the whole subsequent life history of the organism and results in tissue-specific patterns of gene activity and inactivity. Where imprinting discriminates between the maternal and paternal alleles at a single locus, developmental chromosome marking discriminates between different loci. While the context sensitivity of imprinting only entails distinguishing between male (spermatogenesis) and female (oogenesis), the chromosome marking that is part and parcel of nuclear differentiation must be responsive to many features of a cell's milieu.

The role of chromosome marking as a stable parallel system of biological "information" can be observed in a number of ways. High levels of methylation are found in discrete regions of otherwise transcriptionally active "euchromatin," in the large expanses of transcriptionally inactive "heterochromatin," and throughout the entire inactivated X chromsomes of females.[4] These patterns of inactivation are passed on somatically through the course of the organism's lifetime (Jablonka & Lamb 1995). As inductive interactions during ontogeny result in cellular differentiation, tissue-specific patterns of methylation are seen to play a role in allowing differentiated cells to give rise to entire lineages which inherit and retain the differentiated pattern of gene expression. Changes in patterns of methylation have also been implicated as part of the adaptive response of cells/organisms to environmental challenges. While the mechanisms of heritability of chromatin marking are fairly well characterized, the means by which de novo marking is contextually induced is not well understood.

As a heritable "parallel" system of biological order, chromatin marking is deemed to be of evolutionary significance. CpG methylation is an ancient structure found in one-celled organisms whereby the ability of the cell to create epialleles in response to environmental signals provides in effect an environmentally inducible system of variation. Perhaps more importantly, the ability of cells to induce patterns of chromatin marking in each other may have been a key to the evolution of multicellularity. The gambit of multicellularity involves a balancing act between the potential benefits of specialization with the danger that specialized parts will compete with one another. The successful multicellular individual has achieved the ability to straddle that line. The complex, highly differentiated, multicellular organism achieves its system-level integration through the course of a developmental life history. Whereas most plants and certain invertebrates can be regenerated from an aggregate of tissue—i.e., the interplay between the cell lineages represented in the aggregate can lead to the reproduction of the larger developmental pattern—in those organisms deemed to be most complex, this is not the case. Where the challenge of balancing differentiation and integration is greatest, it appears that the replication of the whole developmental life history is required. The organism must be reproduced from the one-celled stage with a concommitant erasure of the epigenetic memory of the previous generation. It may thus be the case that the achievement of an advanced level of specialization is protected by the inability of any of the differentiated parts to give rise to a new organism that lacks the competence to reproduce the whole repertoire of subspecialties.

For the vast majority of extinct and extant organisms (protoctists, fungi, plants, most invertebrates) that do not sequester their germ lines early in development, the production of heritable epialleles through epigentic chromatin marking is likely to have had direct evolutionary significance. It may provide a key mechanism for the genetic assimilation of phenotypic adaptations. To what extent adaptive "marks" influenced by the life histories of parental organisms may also be represented in vertebrate imprinting, is yet to be determined. The further elucidation of the processes of chromosome marking will provide another important window on the plasticity and contextual responsivness of the nuclear constituents of the cell.

From the Hereditary Code-Script to Cycles of Contingency

Fifty years ago Schrödinger argued that the kind of specificity of order which underlies the faithful inheritance of a characteristic such as the "Hapsburg lip" must rely upon a new principle of order-from-order. He then based his influential paean to the hereditary code-script on the claim that only solid-state covalent bonds could be the ultimate source of such order. I've endeavored to "take Schrödinger seriously" by reconsidering, with the benefit of 50 years of hindsight, the empirical basis for this latter claim. I have offered evidence and argument on behalf of the view that biological order is realized, preserved and propagated on many mutually dependent levels. Just as the roots and shoots (foliage) of a tree are circularly the cause and effect of each other (to recall the language of Kant), so too are the multiple aspects of biological order—invested in membrane structure, dynamic regimes, nucleic acid sequence and nucleic acid modification—the cause and effect of each other. Biological order reposes upon highly regulated, energy dependent membrane flow just as much as it does upon genetic sequence stability. Perhaps even more to the point, the very idea of order from crystalline stability must be reworked into an idea of order from dynamic steady-state systems within which crystalline structures function as *constituent resources* that are subject to dynamic modifications, rearrangements, expansions, contractions, and replications. Fifty years of hindsight reveals that life is indeed an order-from-order phenomenon but rejects the claim that the chemistry of the solid state provides the only, or even a privileged, basis for this order. With the rejection of the warrant for any claims for the primacy of a hereditary code-script, the door opens for a new overarching perspective on the nature of the living organism. And just such a perspective is beginning to emerge.

Drawing on a number of disciplines and contributions [including an earlier expression of this work (Moss 1992)], new advocates of a developmental systems theory (DST) are beginning to explore the implications of a biology which is neither explicitly nor implicitly encased in the metaphorical space of the code-script.

From a DST perspective, ontogeny is best viewed as contingent cycles of interaction amongst a heterogeneous set of developmental resources, no one of which 'controls' or 'programs' the process. These resources range from DNA to cellular and organismic structure and to social and ecological interactions. Many of these resources, both inside and outside the organism, can be reliably reconstructed down evolutionary lineages. Evolution is change in these developmental cycles. The changes in gene frequency often used to define evolution are but one aspect of the richer complex of stabilities and changes captured by the developmental systems approach. (Oyama, Griffiths & Gray[5] 2001)

From the DST perspective (Oyama 1985, Griffiths & Gray 1994, Griffiths & Gray 2001) the achievement of any phenotype will rely on the presence of some set of heterogeneous resources, none of which singly determines it and the absence of any of which—be it a vitamin, a gene, or some other developmental cue—may equally result in a characteristic aberration. The developmental systems perspective is thus permissive of many different kinds of biological explanations which may be tailored to local needs and local contexts. At minimum it provides a perspectival antidote to that malady by which the richness and vitality of life processes are lost by slippage into that which is (genetically) known in advance. If no developmental resource is necessarily accorded causal primacy then any explanatory account must place all its cards on the table. When the conflationary temptation to pack contingent developmental outcomes into the one-dimensional sequence arrays of polymers is set aside, then the coding sequences of DNA may be reconceptualized as one type of resource among many, i.e., as Gene-D (Moss 2001). Beyond just a formal parity of resources, DST provides a new opening within which the relationship between different hierarchical orders of the living world may be reconceptualized.

4

Dialectics of Disorder: Normalization and Pathology as Process

Epigenetics (Waddington's term for the analytical study of individual development) can shed light on the process of tumorigenesis, and vice versa.
—Julian Huxley, 1958

Overgrowth or dedifferentiation are effects of . . . disorganization—repercussions not driving forces. Cancer is no more a disease of cells than a traffic jam is a disease of cars. A lifetime of study of the internal-combustion engine would not help anyone to understand our traffic problems. The causes of congestion can be many. A traffic jam is due to a failure of the normal relationship between driven cars and their environment and can occur whether they themselves are running normally or not . . . Cancer is a disease of organization, not a disease of cells . . .

What we need most at present is to develop an autonomous science of organismal organization, the social science of the human body; a science not so naïve as to suppose that its units, when isolated, will behave exactly as they do in the context of the whole of which they form a part, and willing to recognize that whole functioning organisms are its proper concern. I will try to explain normal growth, differentiation, maintenance, and repair, as well as their disorder. It will take biological orderliness in action as its field of study. It lies, in wait for a name, between cytology and sociology. It is much more than oncology, for it is the study of the organization of whole organisms as well as that of disorganisational tumour formation.
—D. W. Smithers, 1962

It is well known from classic genetics that the expression of any multigenic phenomenon is very dependent on the genotypic milieu, so that a given mutation may be deleterious in one genetic milieu and advantageous in another. Thus, the combination of mutagenic changes in genotypic milieus which are different in every human, plus the sensitivity of multigenic phenotypes to the surrounding environment, account for the difficulty in predicting the likelihood of non-familial or sporadic cancers or their outcome once they appear. . . . Even where there is a dominant germ line mutation that favors development of cancer with a probability approaching unity, the time of onset cannot be predicted, and only a very small fraction of the cells, all of which carry the mutation, become transformed; to do so, additional mutations are required, but they can be found in normal tissue as well. To achieve a better understanding of cancer, it will be necessary to take into account the genome of the transformed cell, the state of the surrounding tissue, the age of the organism, its diet and the environment in which it lives.

—Harry Rubin, 1999

From Black Bile to Misguided Developmental Potential

The history of the biology of cancer can be divided into three periods. The debate between a genetic versus an epigenetic-developmental emphasis in explaining cancer can only be dated as far back as Boveri's somatic mutation hypothesis of the first decade of the twentieth century. The commencement of the genetic position in cancer thus squares well with the phylogenetic turn in biology (see chapter 1) and marks the beginning of the third period. The developmental view first emerged during the nineteenth century alongside, and in direct relation to, the rise of modern cell theory, histology, and embryology, all of which had their origins in the ferment of the 1790s. We will thus identify the last decade of the eighteenth century as the beginning of the second period.

The first period of cancer biology begins with the Greeks. Classical medicine referred to cancer but did not distinguish sharply between inflammation, ulcers, benign lesions, and neoplasia—all were taken up in a humoral theory of disease (Rather 1978). Hippocratic writers associated the properties of the four elements, hot, cold, wet, and dry, with the four humors, blood, phlegm, black bile, and yellow bile. In some manner, the historical details of which are not clear, blood became associated with hot and moist, phlegm with cold and moist, yellow bile with hot and dry, and black bile with cold and dry.

This scheme, appropriated by Galen, became canonical for medicine well into the seventeenth century (Rather 1978). Good health in Galenic medicine involved maintaining the right balance of the humors. The source of the humors was understood to be ingested food, which was broken down through "concoction" and distributed through the body.

The general approach to all inflammations was one of understanding them in terms of humoral flows. A tumor was understood to be the result of the damming up of humoral flow with a localized buildup of humors. A flux of black bile mixed with blood gave rise to a kind of inflammation Galen called "scirrhus," which in some cases resulted in cancer. A flux of black bile unmixed with blood gave rise to cancer directly. While cancer could arise in any part of the body, the most familiar case, and the source of the classical identification of the disease with a crablike morphology (hence the name cancer), was that of the female breast (Rather 1978). With Harvey's seventeenth-century work came a shift in understanding toward the recognition of the circular flow of blood powered by the pumping of the heart in place of the classical notion of the expulsive and retentive forces of organs—yet the humoral model of cancer was otherwise largely retained.

During the nineteenth century, much debate turned in relation to the question of whether the orgins of cancer were systemic and constitutional or localized in nature. This dispute is one which has reappeared in different dress ever since. The humoral theory is the classic case of a constitutional (diathesis) view. It is constitutional because the source of the disease it describes is systemic. A tumor may appear at a certain location, but it is the state of the whole body that is responsible for the humoral imbalance (dyskrasia), thus causing the cancer. The loss of support for the humoral theory by the end of the eighteenth century marks the end of the first period of cancer biology but not the end of constitutional theories of cancer. Another constitutional theory of cancer was upheld, for example, by Paget, who in 1853 attributed cancer to two factors: a morbid material circulating systemically in the vascular system and an inherited predisposition to the reception of this impetus toward cancer (Triolo 1965)—a view which resembles current notions of "genetic susceptibility."

A localized etiology of cancer can be based either upon an external cause that acts locally or a discrete endogenous cause. Only the former was entertained toward the end of the eighteenth century. Based on studies of scrotal cancer in chimney sweeps, Percival Pott (1775) (Haggard & Smith 1938) introduced the idea that cancer could be the result of environmental influence.[1] In 1776, Bernard Peyrilhe, following suit, suggested that cancer was a local process, which when presenting diffusely did so by having spread through the lymphatics. Despite Pott and Peyrilhe, the predominant view until the 1830s was still that of cancer as a constitutional disease with localized expression in the form of clotted and degenerated lymph (Haggard & Smith 1938).

The idea that cancer could be based upon an endogenous, local cause was a direct outgrowth of those late seventeenth and early eighteenth century advances in histology, cytology, and embryology which mark the beginning of the second period and the commencement of modern biology. The central figure and acknowledged doyen of this new biology of cancer was Johannes Müller, whose neo-Kantian "teleomechanist" approach to developmental morphology was described in chapter 1. Cancer can be viewed as a local and endogenous aberration when it is analyzed in terms of following the developmental epigenesis of the organism from (1) germs to germ layers, from (2) germ layers to tissues and organs, and from (3) tissues and organs to whole systems. It is again worth considering—but now from the angle of cancer biology—the path by which a neo-Kantian research logic led to Müller's pioneering work in histopathology, and its canonization by Virchow.

The extension of a teleomechanist program can be seen in the work of Blumenbach's student Christoph Girtanner. In his 1796 publication *Über das Kantische Prinzip für die Naturgeschichte*, he "introduced the notion of the Stammgattung, defined as a generative stock of *Keime und Anlagen* which determined certain limits of structural adaptation and which under appropriate environmental conditions became manifest as different but related species (Gattungen)" (Lenoir 1982).

Carl Friedrich Kielmeyer, who studied in Göttingen from 1786 to 1788, gave an influential series of lectures in the 1790s in which, drawing not just upon comparative anatomy but also animal ethology, pathology, and paleontology, he attempted to adumbrate a research program oriented toward a general theory of animal organization (Physik des Thierreiches). Most importantly he introduced the idea of analyzing the unity of the generative stock of *Keime und Anlagen* through comparative embryology. Examining the patterns of embryological development, Kielmeyer envisoned an opportunity to elucidate the dynamic path of animal organization itself. He imagined paths of development to be the source of phylogenetic innovation, but like Kant and Blumenbach, he did not assume a full chain of Being.

Many species have apparently emerged from other species, just as the butterfly emerges from the caterpillar . . . They were originally developmental stages and only later achieved the rank of independent species; they are transformed developmental stages (Lenoir 1982, p. 43).

The beginnings of a modern theory of embryonic germ layers can be traced to Ignaz Döllinger who, although influenced early in his career by Schelling, turned away from a transcendental approach and toward the Göttingen School. Döllinger defined physiology as consisting of two principal fields of study: morphology, which is the general study of organic form, and histology (Lenoir 1982). The germ layer concept was introduced in 1817 by Döllinger's student Pander (Rather 1978) and further generalized by another student, Karl Ernst von Baer. Along with Johannes Müller, von Baer became the seminal figure in the formulation of a developmental morphology structured by the teleomechanist outlook. Von Baer, like many other young Germans early in the nineteenth century, was also influenced by the eminent French anatomist Georges Cuvier.

Cuvier was not from the teleomechanist tradition but shared with it a desire to find the laws of organic form through a comparison of taxa guided by a regard for the functional unity of the whole organism. He was thus an adamant opponent of the romantic *Naturphilosophen*, who sought to understand morphology on the basis of isolated organs subjected to a kind of geometrical intuitionism. Transcendental morphologists attempted to order a continuous chain of taxa on the basis of geometrical transformations of an individual organ. Cuvier, by contrast, was a functionalist. He analyzed animal taxa with an eye to understanding the functional priority of the whole organism and on this basis grouped all animals into four basic categories, or "embranchments." There was no contunuity between these embranchments, each represented a unique and distinctive space of mophological possibility and permutations within each embranchment (that is, the differences between component taxa) were understood to reflect functional requirements.

Von Baer accepted Cuvier's classificatory framework and imported it into the research program of a teleomechanistically guided developmental morphology.

What prefigured a type, or embranchment, for von Baer was a common stock of *Keime und Anlagen* contained in the germ. Taxa within an embranchment varied according to the extent of differentiation of the organs. A more differentiated organ developmentally passed through all the prior stages of less differentiated organs. Von Baer depicted the process of embryological development within an embranchment as a series of radiations from out of a central node of greatest potential. Each path of differentiation leads through new nodes, each of also constitutes a *Stamm* of *Keime und Anlagen*, but a smaller *Stamm*. Such a developmental node might be common to the embryology of a whole taxonomic subgroup (e.g., a "class") of organisms. From out of this node then would be several paths of greater differentiation leading to nodes which are common to smaller taxonomic categories ("orders," then "families," and so on—see a detailed discussion of this in Chapter 1). The kinship of members of an embranchment can be empirically verified on the basis of this model. If two species are members of a common order, then their pattern of development should be seen to coincide up to that point which represents the *Kieme und Analgen* of that order. Subsequently, their developmental patterns should be seen to diverge. The nature of the divergence (that is, why some organs further differentiate and others do not) should be explicable in terms of functional requirements—the specific adaptative needs of their mode of life.

What determines the degree of differentiation is the functional context of the whole organism such that the level of differentiation of one organ will be related to the developmental degree of differentiation of other sets of organ *Anlagen*. Unlike the transcendentalists who would group taxa along the continuum of a single organ, von Baer would expect certain organs of one taxon to be more differentiated and others less differentiated than those of related taxa, depending on the nature of the functional adaptations of the respective organisms. Von Baer characterized the vertebrate type by five fundamental organs out of which all other organs were understood to be derived.

Each and every organ is a modified part of a more general organ, and in this respect we might say that each organ is already contained in all its specificity in the fundamental organ . . . The respiratory apparatus is a further development of an originally small part of the mucous tube (Lenoir 1982, chap 2).

The fundamental vertebrate organs, in turn, are derived from the germ layers. Building on Pander's idea of germ layers, von Baer distinguished two primary divisions of the germ: an upper which he called animal and a lower which he called the vegetative, or mucous, layer. From the upper layer come the sensory organs, skeleton, muscle, and nerves. From the lower come the mesentary organs and digestive and vascular tissue. A middle, or mesoderm, germ layer was first identified by Müller's student Remak in 1850 (Lenoir 1982).

Johannes Müller met Heinrich Rathke and von Baer for the first time in 1828 at the *Versammlung Deutscher Naturforscher und Ärtzte,* organized by Alexander von Humboldt in Berlin. Von Baer was asked to demonstrate the existence of the mammalian ovum for them (described in chapter 1). Müller reported that he was deeply impressed by this demonstration. This experience and the discussion with von Baer and Rathke led, upon Müller's return, to his own work on the Wolffian body, which was published in 1830 and dedicated to Rathke. Müller's seminal work on the human urogenital system constituted one of clearest examples of the power a "teleomechanistic" developmental morphology.

Müller, after his own early brush with the *Naturphilosophen,* became a lifelong opponent of both transcendentalism and mechanistic reductionism. He was ever wary of any investigative methodology that disrupted the unity of the organism. Müller adopted from von Baer the view that the members of a common type hold the same fund of structural elements, with differences in organ formation being a function of different grades of differentiation along a common pathway. Another closely related assumption was that any claim of a causal relation between an earlier and later structure had to be established by a sequence of observable structural transformations capable of being shown to be materially interconnected with one another. This doctine was exemplified by his elucidation of the developmental pathway of the female urogenital system described in chapter 1.

Müller published two papers on tumor microscopy as early as 1836. Developmental morphology already provided the basis for a tumor histology, but it took the achievements of cell theory to provide for the beginnings of a truly modern cancer biology. The formation of cell theory by Müller's students Schleiden and Schwann began in 1838, and its quick appropriation by Müller for cancer theory must also be understood in the context of the teleomechanist outlook.

Although Müller referred to "cells" in his 1836 papers, he intended little more by this word than it had denoted since the time of classical medicine, that is, not a living entity but merely a potential space as, for example, in reference to the cells of a honeycomb. Such was also the sense of Robert Hooke, who looked microscopically at both organic and inorganic specimens, finding cellular arrangements in both sets of cases (Rather, Rather et al. 1986). While the notion of a cell as a living entity doesn't come on the scene until the early 19th century, the concept of fibers as basic constituents of organic tissue dates back to Aristotle and Galen.

Galen posited three basic fiber types. His categorization largely held sway until Giorgio Baglivi (1668–1707) examined macerated animal tissue under the microscope and settled on two basic fiber types (Rather, Rather et al. 1986). Hermann Boerhaave (1668–1738) pushed fiber theory in the direction of a single hollow structure deemed to be filled with a tenuous fluid or spirit. Fibers construed as containers can be seen as a kind of middle ground between classical fibers and the idea of cells. It is worth noting that although both simple and compound microscopes had been in use since the middle of the seventeenth century, they did not have much influence on what was perceived. Unlike the case of astronomy where the telescope largely expanded the range of a *familiar* object domain, in biology microscopes opened up whole new worlds and thus crises of interpretation. The analysis of that which appeared de novo under the microscope awaited new criteria for which objects should be counted as real and important and which should be ignored as ephemeral or artifactual (Rather 1978). New findings were met with much skepticism. Some of the so-called globules reported by early microscopists indeed are likely to have been bona fide cells.

The first clear move in the direction of modern cell theory was made by Johann Christian Reil, whose 1795 essay on the *Lebenskraft* (life-force) postulated the crystallization of fibers into a living form based on the presence of a nucleating germ (Rather, Rather et al. 1986). Reil, who had also studied in Göttingen between 1779 and 1789, overlapping with Blumenbach, had conducted his own appropriation of Kantian ideas for biology. Reil pursued a theory of a *Lebenskraft* based on the idea of chemical affinities. The specificity of a *Lebenskraft*, which was for Reil the basis of the specificity of a species, was constituted by the arrangement of chemical affinities in the *Kern* or *Stamm*, an arrangement which was passed from one generation to the next by way of the egg (Lenoir 1982). The idea of an egg as the repository of a developmental potential embedded in the organization of germinal particles went on to play a crucial role in the conceptualization of cell theory.

As recounted in chapter 1, modern cell theory proper began in Müller's laboratory in the 1830s with Schleiden's work on plant cells, quickly followed by Schwann's extension of cell theory to animal cells. Within months of Schwann's generalized exposition of animal cell theory in 1838 and with the aid of the newly invented achromatic microscope, Müller reported finding cellular structure in a pathological growth in which he had failed to see it before. The cell theory provided Müller with a powerful basis for the classification of pathologies using the analysis of cell type and tissue development. With cartilage as his exemplar, Müller found that the hallmark of pathology is an arrested or aberrant path of differentiation, in both cases an expression of cellular potential which has become separated from the *principle of the whole organism*.

The differences between pathological and healthy cartilaginous developments consists principally in the continuation of embryonic cell formation. In numerous other tumors the same observation can be easily made. It is not the form of the elementary parts that distinguishes diseased structures. The problem lies in part in the formation of normally primitive structures where they are not necessary and do not contribute to the purposiveness of the whole, and particularly in the incomplete development of these tissues, which usually only reach a particular stage of development that is transient in healthy life. This is the mode of operation of diseased vegetative life. In the development of sound primitive cartilage, however, the monadic life of the cells is controlled by the *Lebensprincip* of the entire individual; it reaches its limit, the cells coagulate and the interstitial, unclear fibrous mass of the cartilage emerges between the cavities of the germinal cells. In the Enchondroma on the other hand the regulated life of the part no longer attains a particular limit, and it slowly continues to increase in size. The cell walls in this case do not thicken normally; the cartilage remains in its embryonic condition and this embryonic structure is continually repeated. *Über die krankhaften Geschwülste* (in Lenoir 1982, pp. 144–145).

With the conceptualization of the cell as the now more specified repository of the *Keime und Anlagen*, development and cancer emerge as in effect complementary possibilities in a story about the relationship of monadic parts to that organismic whole that is constituted by them. In this view, far removed from classic notions of humoral-based inflammations, we can see a model of cancer recognizable in its antecedent relationship to strains of current thought. It was through Müller's teleo-mechanistically informed reflections upon the relationship of parts to whole, in which the parts possess the potential of the whole, yet a potential whose realization is mediated by the interaction of the parts under the auspices of the whole, that such a model takes shape. Against the practices of more simple-minded empirical contemporaries, Rudoph Virchow, student of Müller and heir to cellular pathology, criticized the practice of ontologizing pathological structures. He suggested that Müller's greatest contribution was that of understanding the "law of the identity of embryonal and pathological development" and its corollary that histopathological lesions of "pathological products" should not be considered as "given, ontologically complete things but merely as tissues in stages of development" (Rather 1978).

It was not until the 1850s that Virchow and a fellow former Müller student named Remak could definitively rule out the possibility that cells arise not only from other cells but also from acellular structures—"cytoblastema." Having established his famous dictum that *Omnis cellula a cellula* (cells come only from cells), Virchow could then proceed to classify all tissues in cellular terms. Virchow (1858) placed all tissues into three groups: those with cells adjacent to one another (epithelial), those with cells separated by intercellular substance (connective), and those which have undergone a further transformation: muscle fibers, nerve fibers, blood vessels, and so forth (Rather 1978). Where the former cytoblastema theory had posited a unique germinal layer from which all cancers as well as new growths emerged, the move to the doctrine of cellular continuity meant that the origins of cancer could and must be localized to a specific tissue and cell type. At this point Remak and Virchow diverged. Virchow, who had been the first to identify the cellular basis of stromal (connective) tissue, postulated stromal tissue to be the principal source of undifferentiated cells that could then be transformed into epithelioid tumors. Remak, by contrast, held to the epithelial origin of cancers (Triolo 1965).

The Bonn anatomist Franz Boll (1876) made an attempt to mediate the conflicting claims of stromal versus epithelial priority in cancer causation. He put forward a theory of cancer based on tissue interaction. Retracing Remak's studies on lung differentiation in the embryonal chick, Boll suggested that normal development entailed a history of appositional "conflict" between germinal connective and germinal epithelial tissues (Triolo 1965). In this model the formation of a normal lung was not the result of a unilateral growth principle but rather a compromise between two dissimilar though reciprocally determined principles. Bonn added a new dimension to the vision of epigenesis, that of inductive interactions. Yet one can see that this possibility was already implicit in the move to a cellular histology in which the *Keime und Anlagen* of the whole had become disseminated into monadic parts. For Boll, the cause of abnormal growth was the *untimely* resumption of interface antagonisms prompted by exposure to external irritants. Boll thus made the first move toward explaining, in dynamic epigenetic terms, the effects of environmental carcinogens. His views bear some resemblance to the epigenetic model of carcinogesis put forward by Farber and Rubin (to be discussed in detail below) a century later.

Within the same logical space as Müller's anomalous germ cell, Virchow's idea of a plastic embryonal-like connective tissue stratum, and Boll's formative tissue interactions was Julius Cohnheim's 1875 notion that cancer was derived from an embryonic residue that retained its full potential. Cohnheim conceived of an overproduction of cells occurring during organ morphogenesis. Some of these cells, he hypothesized, detached prior to completion of differentiation, thus retaining the ability to re-emerge with an active capacity for proliferation at some later date (Triolo 1965).

While this model proved to constitute an anticipation of the embolus theory of cancer metastasis, it suffered on several counts as a basic theory of cancer origination. Specifically, it failed to locate the presence of residual embryonic cells. It also failed to account for why such cells, if present, would begin rapid growth at one time versus another (Rather 1978). A similar yet independent model had been put forward by the Italian surgeon Durante (1874), who identified "embryonic rests" with nevi (birthmarks). He envisioned the effects of some kind of irritant in triggering a renewed proliferative capacity of "elements which have retained their anatomic embryonic characters in the adult organism" (Triolo 1965). Beginning in 1894 the histopathologist Hugo Ribbert hoped to answer criticisms of Cohnheim's model by presenting a more circumspect theory in its place. Ribbert suggested that normal growth was due to the "coordinated and balanced" development of its cellular components. If this tissue "tension" were lost, and in direct proportion to the extent to which it was lost, deviation from normal processes would then ensue (Triolo 1965).

Oncology after the Phylogenetic Turn

The legacy of the nineteenth century for twentieth-century oncology was a framework for investigating cancer in terms of a dynamic relationship of monadic cellular parts to the organismic whole of which they are constitutive. From Virchow through Boll, Durante, Cohnheim, and Ribbert one can see different attempts at modeling carcinogensis within a common framework. There is of course no single moment at which point twentieth-century oncology takes its gene-centered, phylogenetic turn, and certainly no research finding in oncology that leads it in this direction, and yet the shift in interpretive orientation is very clear. Perhaps owing to the lack of a research exemplar that compellingly illustrates the productivity of the phylogenetic perspective in cancer biology, the monadic-part-to-whole-composed-of-monads perspective is never fully lost; rather it is maintained as an ongoing undercurrent which is periodically rediscovered and rearticulated. What does occur is a bifurcation such that the perspectives of Müller, Virchow, Cohnheim, et al. lead to two fundamentally different ways of interpreting the nature of the cell that becomes aberrant.

Along with other areas of biology (see Allen 1975; Harwood 1993), cancer research in the twentieth century became oriented toward experiment and the establishment of good experimental models. Experience with exposure to industrial soots, tars, and dyestuffs during the last two decades of the nineteenth century provided empirical support with public health motivation for an "irritation hypothesis" of cancer causation. The theories of Virchow, Thiersch, Cohnheim, and Ribbert were used by experimentalists seeking to establish explanatory models for irritant-induced carcinogenesis (Triolo 1965). One of the first model systems of cancer causation was established by Katsusaburo Yamagiwa, a student of Virchow, who had returned to Japan. He had succeeded in his efforts to experimentally produce skin cancer through painting tar onto the ears of rabbits. It was exactly at the point of interpreting the process of experimentally induced carcinogenesis that a bifurcation of explanatory perspectives took place.

In the holistic framework of Virchow et al., the monad-like cell is invested with the potential to develop in any number of ways. What emerges from this legacy as the *common* focal point for twentieth-century thinking about cancer is the notion of the "autonomous cell" (Triolo 1964). But company parts with repect to the meaning of an autonomous cell. To heirs of nineteenth-century holism, autonomy was understood in terms of "totipotency," the possession by the cell of the potential of the whole. The autonomy of the cell understood this way is then the precondition for either normal or aberrant growth and a prior guarantee of neither. What determines which way it will go, normal or aberrant, is not its internal features but the subsequent history of its interactions. Thus the potential danger of the autonomous cell becoming cancerous is the flip side of its adaptive potential. This is a point that will be amplified later. From a certain viewpoint consistent with the developmental-holistic perspective, cancer is an adaptation, albeit one in which the "transformed" cell has become uncoupled from the developmental matrix of the organism. It is adaptive behavior at the level of the cell and its progeny but not at the level of the organism for which it is anything but adaptive.

The second interpretation of the meaning of the autonomous cell is that which first arises in the early twentieth century. Here cellular autonomy is no longer perceived as the normal state of affairs that is the precondition of either normal or abnormal growth but rather as a distinctively aberrant condition that is synonymous with cancer. Cancer is no longer conceived as the product of the contingent interaction between cells and other cells and between cells and their extracellular environment but rather as determined from within a cell. This model of cancer is one in which cells from somatic tissue become autonomous due to a mutation.

The first somatic mutation hypothesis was credited to Boveri by virtue of his publication of "Zur Frage der Enstehung maligne" (1914). He based his ideas on the earlier cytological studies of von Hansemann on chromosomal irregularities associated with cancer. These particular irregularities, however, were also found to occur in the absence of carcinogenesis. Boveri himself never observed tumors cytogenetically. Nonetheless, he attributed cancer to the formation of an aberrant chromatin complex.

The *somatic mutation hypothesis* was a direct expression of that gene-centered pattern of interpretation that I have also referred to as the "phylogenetic turn." The combination of Mendel's laws and de Vries's mutation theory provided the theoretical foundation of the somatic mutation hypothesis. It was additionally revised by Bauer in 1928 in order to accommodate H. J. Muller's work on inducing mutations in *Drosophilia* by X-ray exposure (Lawley 1994). As genetic research increasingly took center stage among twentieth-century American biologists, the holistic-developmental understanding of cellular autonomy became marginalized.

Animal studies in experimental carcinogensis, which presupposed only the somatic mutation model, emerged as significant research programs. While such studies established the ability to induce experimental cancers with carcinogens in the laboratory, they did not necessarily lend empirical support to the somatic mutation hypothesis. Data that would arguably militate against the somatic mutation model were also recorded. In some of the earliest studies in experimental carcinogensis (1906), a German pathologist found that a dye, scharlach R, when dissolved in olive oil and injected beneath rabbit epidermis, would cause cells to behave as if malignant. This finding of the production of an apparent squamous cell carcinoma was readily reproducible by other investigators, but rather than being an irreversible processes, consistent with a mutational hypothesis, the cells reverted to a normal state as the influence of the dye wore off (Rous 1959). Contrary findings such as this, however, proved to be no match for the growing gene-centered, antidevelopmental momentum.

What was needed in order to *harden* the somatic mutation hypothesis was a reliable correlation between carcinogenesis and mutagenesis, but this proved to be elusive. No chemical mutagen was well established until after World War II, and because the one that had emerged—mustard gas—was used as an agent of chemical warfare, publication of any findings pertaining to it had been forcibly delayed. Prior to the availability of mustard gas the only established means for inducing mutation was through X-ray exposure, a method discovered by Muller and applied to *Drosophilia* genetics beginning in the 1920s. Muller, however, did not begin to apply this to cancer research until the 1950s. At that time he observed that "cancers induced by overexposure of a part of the body to radiation often fail to show up until some 10 to 20 years after the cessation of the irradiation" (Lawley 1994).

In addition to the experience of latency in X-ray induced cancer, observations of a strong correlation between advanced age and cancer incidence led to a "multi-hit" theory of cancer causation. In this model cancer is never the result of a single somatic mutation; rather, some number of sequential mutations are required for carcinogenesis. Initially, Muller and others proposed that as many as six mutations were necessary. But the problem with that theory is that spontaneous somatic mutation is understood to be a rare event. In order for six sequential mutations to accumulate there would have to be an adequate amount of time for each newly mutated cell to expand clonally, that is, give rise to new generations of genetically identical cells such that there were enough second generation cells (with the first mutation) that the chance of one of these acquiring another mutation was higher than negligible. The time frame required for five such sufficiently large clonal expansions, however, appeared to be inconsistent with the possibility of cancer's occurring within a human lifetime. This caused the cancer research community to shift down to a two-hit model and yet "it became evident from studies of the development of specific types of cancer that often more than two genetic changes are involved" (Lawley 1994).

Proponents of the somatic mutation theory of multistep carcinogenesis had adopted the nomenclature suggested by Peyton Rous during the 1940s. Rous distinguished between initiation, which was a necessary but not sufficient precondition for carcinogenesis, and promotion, which, given an initiated state, could end the latency period, resulting in the beginning of phenotypic transformation and finally the progressive malignant transformation of the cancerous tissue. Rous himself, however, was never convinced by the somatic mutation hypothesis. In his studies of 1941 he found the initiation of tumors in skin painted with tar to be reversible. If skin painting was interrupted, the tumors disappeared but could be induced to reappear at precisely the same sites with renewed application. Additionally, he found that the priming, or initiating, effect could also be achieved through the use of noncarcinogenic stimuli, such as the application of turpentine or the process of wound healing in response to the boring of holes in rabbit ears (Lawley 1994). With words highly reminiscent of the irritation hypothesis of Virchow et al., Rous observed that:

The great majority of carcinogens are merely provocative: they convert normal cells into tumor cells but have no further essential role and are degraded or left behind as the changed cells multiply. Their relation to the neoplastic state may be likened to that of ignition to combustion: a fire can be kindled in any one of numerous ways but, with this done, its decisive share in events is ended. The analogy can be pushed further: flames differ according to their ingredients and tumors differ according to the type of cell involved (Rous 1959).

Addressing the somatic hypothesis more directly Rous proclaimed that "the somatic mutation hypothesis, after more than half a century, remains an analogy . . ." (Rous 1959). Nor was Rous alone in his assessments. In his comprehensive review of the somatic mutation hypothesis Burdette (1955) concluded that "a general correlation between mutagenicity and carcinogenicity cannot be proposed from present evidence." This opinion was echoed nearly 15 years later by Leslie Foulds in his authoritative review of the state of the art (Foulds 1969).

Over a half a century of studies indicated that some of the most carcinogenic agents were not mutagenic, and that some of the most mutagenic agents were not carcinogenic. Rous, while characterizing the effects of carcinogens as being merely provocative, having no influence on the nature of the cancer that resulted, offered by contrast evidence of "neoplastic viruses" which did appear to influence the nature of the cancer that followed. He suggested that rather than somatic mutations "the neoplastic effects of the neoplastic viruses may be due to their own activity which is reproduced and transmitted" (Rous 1959).

Even cancer biology can have its ironies. The somatic mutation hypothesis, as we will see, reached its high water mark, not by use of X-ray induced mutation or through the use of chemical carcinogenesis but rather by way of studies using a neoplastic virus. And the neoplastic virus used was the very one named in honor of the founding father of tumor virology, Peyton Rous.

Oncogenes and Oncogenesis

The development of the oncogene concept emerged from, and has been closely tied to, the history of tumor virology. The first virus clearly shown to be the cause of tumors in an animal was isolated from chickens by Peyton Rous in 1911. However, this and subsequent findings in tumor virology had been treated as separate areas of research from that of the work in experimental carcinogenesis discussed above. The notion of a tumor etiology from viral infection did not fit in with the prevalent thinking of oncologists. Fowl tumors were typically dismissed as being somehow irrelevant to mammalian, and especially human, cancer (Huxley 1958).

The turn toward neoplastic viruses for insights into a general theory of cancer causation received a fateful boost with the efforts of Harry Rubin, who joined the laboratory of Renato Delbecco at Cal Tech in 1953 in order to establish a quantitative basis for analyzing virally induced tumors (Rubin 1955). With the assistance of then graduate student Howard Temin, Rubin succeeded in establishing a cell-culture system for analyzing the chicken virus he acquired from "Old Man Rous," that would prove to be a springboard for the subsequent development of tumor virology. By the late 1950s several lines of evidence suggested that the Rous sarcoma virus (RSV) was composed of RNA (as opposed to DNA), although the first actual intact isolation and characterization of RNA from the virus was not performed until 1965 (Rubin 1965). This appeared to be paradoxical in light of evidence of the high radiation sensitivity of the cell's capacity to support RSV infection and viral replication, which suggested that like a temperate phage, RSV became integrated into the genome of the host cell (Rubin 1965).

But how could RSV be both composed of RNA and yet become integrated into the host genome? This apparant paradox was resolved with Howard Temin's discovery of the reverse transcriptase enzyme. RSV and other now so-called retroviruses were found to contain, in their RNA, gene-sequence information for the synthesis of an enzyme capable of using viral RNA as a template for the synthesis of complementary DNA, which could then be integrated into the host DNA. In addition to solving the paradox of retroviral RNA composition and radiation sensitivity, the discovery of the reverse transcriptase enzyme (for which Howard Temin along with David Baltimore were awarded Nobel Prizes) constituted the addition of an invaluable tool for biotechnology. With the reverse transcriptase in hand, molecular biologists could synthesize complex sequences of DNA (so-called cDNA) using "messenger RNA" found in cell cytoplasm as templates.

By 1969, with experimental systems for studying tumor virology well established and with the range of taxa in which tumors could be produced by retroviruses expanded from that of just chickens to include also mice, cats, and hamsters, the time had come to bring virally induced, and non-virally induced, causes of cancer into a common explanatory framework. The move toward explanatory unification in the realm of oncology took the form of a series of *oncogene hypotheses*, the first of which was put forward by Huebner and Todero in 1969. Huebner and Todero suggested that cancer, "both spontaneous cancers and those induced by chemical and physical agents," is the result of the expression of viral genes, i.e., "oncogenes" that become resident in the genomes of animals.

The central hypothesis implies, therefore that the cells of many if not all vertebrates carry vertically transmitted (inherited) RNA tumor virus information (virogenes) which serve as an indigenous source of information (oncogenes) which transforms normal cells into tumor cells; additional phenotypic expression of viral information may or may not also occur (Huebner and Todero 1969).

All cancer, in this view, is the result not of any form of inherent potential gone astray but rather of the agency of an enemy lodged within. Such agency is attributed to a kind of infectious, microscopic pathogen which has achieved, for reasons presumed to be explicable in terms of a logic of natural selection, the ability to pass from one generation to the next while entering into tumor-forming activity typically late in the life cycle of the host. This hypothesis introduced the concept of "oncogenes." By conceiving of oncogenes as viral genes lodged in the germ line Huebner and Todero endorsed a model of the cancer cell as *internally determined*, albeit without subscribing to a somatic mutation etiology. Internal determination of an aberrant "autonomous" condition and somatic mutation, in the Huebner-Todero hypothesis, became uncoupled.

In 1971 Howard Temin offered an alternative hypothesis which attempted to account for the origins of the retroviruses and do so in a way which recovered a somatic mutation model of carcinogenesis, albeit with an interesting new twist.

Temin theorized that normal somatic differentiation occurred through a process of sequential somatic mutations facilitated by the activity of the reverse transciptase enzyme. Temin's theory placed carcinogenesis back into the context of the processes of normal development but now with a characteristically "informationistic" orientation.

An organism needs to identify cells in a stable way, so that one cell is identified as a retinal cell at a particular position, and another cell is committed to make antibody to a particular antigen. The most stable storage place for such information appears to be the cellular DNA. RNA → DNA information transfer in somatic cells would provide a mechanism for stable differentiation of DNA (Temin 1971).

Viruses are explained in this view as a contingent side effect of the processes of normal cellular differentiation. In Temin's view, RNA originating from one cell is reverse transcribed into the DNA of another cell, resulting in somatic differentiation. Temin uses the term *protovirus* to refer to that normal sequence of reverse-transcribed DNA, the integration of which into a cell's genome (thus a somatic mutation) causes its proper differentiation.

The process of protovirus transfer might work as follows. A region of DNA in cell A serves as a template for synthesis of an RNA which is transferred to cell B. In cell B, a new DNA is made by an RNA dependent DNA polymerase, using the transferred RNA as template. This new DNA then integrates into the DNA of cell B. This integration could be next to the homologous DNA or at a different place. In either case, cell B would differ from cell A, which remains unchanged (Temin 1971).

Somatic mutation, in Temin's view, is not in itself an aberrant occurrence but rather a basic and unavoidable feature of cellular differentiation. Viruses emerge from such processes strictly by chance. They result when a reverse-transcribed stretch of DNA happens to become integrated into a genome such as to result in a new string of nucleic acids (genes) that are capable of quasi-independent replication. In the vast majority of cases, the formation of the reverse-transcribed DNA (a protovirus) does not result in a virus but only in a properly differentiated somatic cell. Cancer, according to the Temin theory, is a product of the limited fidelity of normal somatic-cell differentiation. Cancer, for Temin, is once again the result of somatic-cell mutation, although only of the odd case of it.

The usual process leading to cancer could be a variation in the normal physio-logical evolution of the protovirus DNA, so that variants which contain infor-mation for the cancer appeared either by mutation of the base sequence or by integration in incorrect places or both (Temin 1971).

Temin's idea of differentiation by somatic mutation, with the notable exception of the immune system, has largely *not* been borne out. Inter-esting as it was, Temin's hypothesis of cancer causation could not be redeemed if directed somatic mutation is not found to be operative in processes of cellular differentiation (other than in the immune system where just such a process is well established).

The kinetics of RSV-induced tumor formation had long been recog-nized as highly variable. Cloning and nucleic acid sequencing studies of the retroviral genome during the 1970s revealed that there were two classes of retroviruses, one which was acutely transforming—capable of rapidly inducing tumor formation in animals—and another which was only weakly transforming, that is, capable of inducing tumors but only after a long latency period. It was soon recognized that these were struc-turally different in only one respect. The acutely transforming viruses were larger than the weakly transforming viruses by one stretch of nucleic acid, that is, by one gene (Cooper 1990). This gene, whatever it happened to be, constituted the difference between acutely transforming and only weakly transforming viruses. Owing to its apparent significance in tumor formation, whatever gene was making the difference was awarded Huebner and Todero's designation *oncogene*.

The first oncogene to be identified, through comparing the nucleic acid sequence composition of acutely transforming and weakly transforming Rous sarcoma virus strains, was named the "src" (for sarcoma) gene (Duesberg 1983). It was found that different so-called species of tumor-causing retroviruses differed from their less-virulent isoforms through single genes, but that all these genes were different. Sequence comparisons of other retroviruses began producing a growing catalog of putative oncogenes. The end of 1982 saw the identification of 17 other such oncogenes from retroviruses (Bishop 1982). An obvious question in the aftermath of both the Huebner-Todero and Temin hypotheses was: Where did the oncogenes come from? Were they of foreign origination and akin to the oncogenes of Huebner and Todero or the stochastic side effect of some otherwise normal process of somatic rearrangement, à la Temin?

Further studies indicated that certain animals that had been infected with a weakly transforming virus eventually expressed a tumor from which acutely transforming virus could then be isolated (Cooper 1990). If a weakly transforming virus enters an animal without an oncogene and then reemerges from the animal with an oncogene it would appear that it "picked-up" the oncogene from within the animal. This much seemed clear. What was left to be determined was whether the putative oncogene picked up from the host was originally derived from a virus, and thus like the oncogenes depicted by the Huebner-Todero model, or was native to the host. And if the latter, then how and when did it become an oncogene?

Working in the laboratories of Harold Varmus and J. Michael Bishop, Dominique Stehlin constructed a molecular probe for the RSV src gene. Nucleic acid probes (DNA or RNA) take advantage of the same chemical features that are used by the cell in DNA replication. Of the four DNA building blocks—A, T, C, G—there is differential recognition and binding between A and T and between C and G. A and T are thus complementary base pairs as are G and C. The principle of nucleic acid polymer replication, whether in vivo or in vitro, is simply that of using a sequence of bases as a template for forming its A-T/G-C complement.

A DNA probe must satisfy two requirements. It must have the correct sequence to bind to the gene of interest (and no unrelated genes) through complementary base pairing with the target DNA (that has been uncoupled from its double helix complement by being separated into single strands). And it must also be linked to some visualizable marker in order to enable the investigator to locate it. When a DNA probe derived from one species is successful at locating a target sequence in another species, it suggests that the target sequence has been highly conserved over evolutionary time and is thus likely to be of much biological significance. In the absence of a need for preserving the specific sequence, it is expected that the identity of DNA bases will mutate at a regular rate.

Localization by complementary binding studies, using a src probe derived from chicken to scan the DNA of a variety of other avian species including the Australian emu, indicated that the src gene was highly conserved in avian phylogeny and thus likely to be of functional significance for the host. The level of interspecies sequence variation, as assessed by molecular hybridization studies, was consistent with the characterization of src as a highly conserved gene (Varmus 1989). In order to gain further confirmation of the nonviral origins of homologous src genes, full sequence analysis would have to be performed (i.e., the gene would have to be cloned and sequenced). The homologous genes found in the genomes of other avian species were referred to as "c-src," for cellular src. Sequence analysis of c-src demonstrated the presence of introns (noncoding intervening sequences), where "endogenous virogenes have the insignia of provirus, being composed of continuous coding domains, flanked by repeated sequences" (Varmus 1989). Introns are the intervening sequences of DNA which are found in eukaryotic genomes but not prokaryotic or viral genomes. Their presence in c-src genes again suggested that c-src was not of viral origin. Identification of the source of the viral src oncogene as a putatively functional host gene led the research program in molecular oncology back to the framework of a somatic mutation model.

We said that the RSV transforming gene is indeed represented in normal cellular DNA, but not in the form proposed by the virogene-oncogene hypothesis. Instead, we argued, the cellular homolog is a normal cellular gene, which is introduced into a retroviral genome in slightly altered form during the genesis of RSV.

Far from being a noxious element lying in wait for a carcinogenic signal, the progenitor of the viral oncogene appeared to have a function valued by organisms, as implied by its conservation during evolution. Since the viral src gene allows RSV to induce tumors, we speculated that its cellular homolog normally influenced those processes gone awry in tumorigenesis, control of cell growth or development (Varmus 1989).

What enabled the retrovirus work to breathe new life into the somatic mutation research program was the idea that these viruses pointed the way to the particular genes which, owing to the nature of their normal function, could become causes of cancer in the event of mutation. Comparison of the nucleic acid sequences of the oncogenes with the established sequence data for the genes of known proteins (Genes-D), suggested that the oncogene products fall into five protein categories. All five of these categories have been associated in some way with growth-related functions. These are the following:

1. Growth factors

2. Growth factor receptors

3. Signal transducers

4. Protein kinases

5. Transcriptional activators

An explanatory model emerged which depicted these classes of proteins as the nodal points of a universal growth regulatory circuit. Lesions at any of these points, according to the theory, could result in carcinogenesis. Appraisal of the significance of this model by its proponents was not modest. In his 1982 review Bishop asserted that

. . . it is the retroviruses that have provided the most coherent and penetrating view of tumorigenesis presently available to us. Three features of retroviruses account for this sentiment. First, the oncogenes of retroviruses have provided our first glimpse of enzymatic mechanisms responsible for neoplastic transformation. Second, the diversity of retrovirus oncogenes has provided a rich set of oncogenic agents whose versatility far exceeds that of DNA tumor virus oncogenes, and whose tumorigenic capacities provide separate experimental models for most major forms of malignancy. Third, oncogenes appear not to be indigenous components of retrovirus genomes, but instead have been transduced from normal genetic loci of the vertebrate hosts in which retroviruses replicate. Moreover, we have reasons to believe that the vertebrates from which retrovirus oncogenes derive may participate in tumorigenesis induced by agents other than viruses. Thus while tracking the evolutionary origins of oncogenes, retrovirologists have been led well beyond the confines of tumor virology, to confront what may be a final common pathway of oncogenesis (Bishop 1982).

In the oncogene model of Huebner and Todero cancer is caused by genes (given the necessary level of activation), but they are genes derived from a foreign source and oriented toward a foreign mission. Cancer would then not represent an intrinsic potential of the biology of the cell-organism but rather a consequence of a foreign agent, the (virally derived) oncogene, which when activated assumes an executive function in the cell. With the recognition that retroviral oncogenes like src are derived from normal host genes which appear to be functional in growth-related activities, a choice had to be made as to how to interpret the manner in which such genes may become involved in a process of carcinogenesis. Either the first or second interpretation of the autonomous cell (discussed above) could have been invoked in relation to the oncogene findings.

Is cancer the original and intrinsic potential of any living cell because it is the unavoidable risk, the organismic downside, of totipotency? Does the capacity to adapt to novel situations—the holistic-developmental sense of autonomy—bring with it a corollary risk of cellular uncoupling and divergence from the organismic hierarchy? In this view genetic mutations in cancinogenesis would in effect be viewed as permissive and not as directive. Mutations would serve to free the cell from those processes which developmentally steer it into a stable pathway of terminal differentiation. Alternatively, is the role of genetic mutations determinative of the cancer phenotype? Does the potential for cancer only first arise after mutations have specified an aberrant autonomous phenotype. Does the mutation code for autonomy? Bishop and Varmus ultimately opted for the latter interpretation, selectively drawing on features of both the Huebner-Todero and the Temin hypotheses.

If cellular proto-oncogenes are normal and functional elements of the cell, then expression of these proto-oncogenes cannot be the cause of cancer (as they would in the Huebner-Todero hypothesis) but rather must first be subject to some form of alteration, that is, somatic mutation (as depicted by Temin). But unlike Temin, Bishop and Varmus did not attempt to situate such mutations in the normal course of development. Carcinogenesis, for Bishop and Varmus, was depicted as the result of the executive-like action of transformed proto-oncogenes (like

Huebner and Todero) which had, in effect, gained a new function for the causation of unregulated growth.

Huebner and Todero's oncogenes were strictly Genes-P. Postulated as the result of the natural selection of viruses, the ascription of a gene for functionality could have followed a Mendelian logic. If the Bishop and Varmus oncogenes could not be the result of any form of natural selection, how could they be accorded the status of a gene for functionality? Yet such status was implicit in the Mendelian (Gene-P) language which Bishop and Varmus cleaved to. "Dominantly-acting oncogenes" became their catch-phrase as well as the veritable motto of the oncogene research program. Bishop and Varmus spliced together pieces of the Huebner-Todero and Temin models in such a manner as to endorse the second sense of cellular autonomy in a big way. Cells were not the repository of adaptive potential, for better or for worse, but rather were depicted as passive vehicles commandeered by phenotype-determining oncogenes. Dominance was invoked in an ambigous double sense—dominant with respect to a nononcogenic allele at the same locus but also dominant with respect to the genes of other loci. The Bishop-Varmus oncogenes, in a sense, oscillated between being Genes-D and Genes-P. They were like Gene-D in the sense that they were sequenced and thus molecular entities. (However, the ability to distinguish clearly at the molecular level between normal cellular proto-oncogenes and activated oncogenes often proved to be elusive.) And they were like Gene-P in being defined according to their phenotypic correlations in transformation assays. There is little doubt that the prospect of being able to unite the criteria for being a Gene-P with the criteria for being a Gene-D (even if this was only tacitly understood) was responsible for generating much of the excitement and fanfare that became associated with the oncogene program.

What would it actually take for the Gene-P and Gene-D criteria to properly coincide in the case of an oncogene, and what would the larger significance of such a coincidence be? Minimally, there would have to be some consistent difference in molecular sequence between proto-oncogenes and activated oncogenes and this difference would have to consistenly map onto a difference in phenotype (of an organism, not just a cell in culture), with the later phenotype being malignant. And yet even if this correlation had been well established, and it never was, it would still not have settled the question as to which sense of the autonomous cell to endorse—how to understand the fundamental nature and cause of malignancy.

Does a mutation in a particular proto-oncogene dictate that a specific malignant phenotype is produced, or rather does said mutation result in a cell becoming uncoupled from its surround and thereby susceptible to embarking on a de novo path of cellular adaptation? In the latter view, malignancy is the result of complex adaptive (autopoietic) behavior which may be triggered by a mutation, but the malignant phenotype will be largely contingent on its response to local conditions.

Cancer by Decree or Cancer by Default?

In contrast to the claims of genetic dominance uttered by the oncogene community, a separate and unrelated program of investigation into the genetic basis of cancer, its chronology paralleling that of the oncogene investigation, appeared to identify a genetically *recessive* basis of carcinogenesis. The concept of tumor suppresser genes originated from two independent lines of work, cytogenetic studies on hybrid cells derived from somatic cell fusion experiments and epidemiological studies on childhood cancers. In both cases evidence accumulated in support of the idea that the genetic lesions associated with the occurrence of cancer were recessive in nature. Cancer was the result of the loss of *both* alleles at a certain locus, and this locus was in some manner associated with constraining cells from entering into unregulated growth.

The idea of the existence of tumor suppressor genes arose, in part, from experiments that involved the fusion of two distinct cells and thus followed from a technical breakthrough. The first evidence of the possibility of viable cellular fusion involving multinucleate cells was made in relation to the observation of pathological tissue by Johannes Müller (Faber 1893). However, the first experimental demonstration of viable cell fusion was not achieved until 1965 (Harris & Watkins 1965). In analyzing the implications of their achievment, the following observations were made (Harris 1970).

1. That an inactivated virus could be used to provide a general method for fusing animal cells together under controlled conditions.

2. That fusion could be induced between cells from widely different species.

3. That the fused cells were viable.

One of the many intriguing avenues of inquiry which this experimental breakthrough enabled was the possibility of fusing normal with malignant cells or transformed cells with different degrees of malignant potential, and then observing the results. In early studies in which cells of greater and lesser malignant potential were fused together and the malignant potential of the resulting hybrids was evaluated, a retention of the more highly malignant phenotype was observed. These results could have lent credence to an interpretation of the cancerous phenotype as an expression of genetic dominance. The results however were deemed to be inconclusive.

In none of these cases, however, were the chromosomes of the tumors analyzed in detail, so that no assessment can be made of the extent to which the results might have been complicated by loss of chromosomes or by selection of atypical variants in vivo (Harris, Miller et al. 1969).

What soon became apparent was that the tetraploid (fours sets of chromosomes) fusion products of cells derived from different species, and with less frequency those derived from animals of the same species have a marked tendency to lose certain chromosomes. Once apprised of this factor the question could be experimentally posed again. The results were surprising. Fusing a variety of tumorigenic mouse cells with cells of a nontumorigenic mouse cell line, Harris and coworkers found that the malignant phenotype was suppressed in *all* cases. In addition, they observed "that hybrids resulting from such fusions produce segregants in which a loss of chromosomes is associated with a reversion to malignancy" (Harris, Miller et al. 1969). The implication of these experiments appeared to be clear. The malignant phenotype is enabled only when all of the chromosomes which carry some factor or factors are lost.

The ability to suppress a malignant phenotype proved to be a widespread phenomenon of somatic cell fusion experiments using both rodent and human fusion partners. The proclivity of tetraploid hybrids to drop chromosomes provided the opportunity to identify the particular chromosomes required for suppressing a malignant phenotype. Where the spontaneous loss of a certain chromosome correlated with the expression of the malignant phenotype, it was surmised that that chromosome, when present, suppressed that malignant phenotype. Mouse chromosome 4, for example, was found to be responsible for suppressing a wide range of tumor cells including those expressing retroviral oncogenes (Harris 1988). Human chromosome 1, which was shown to have significant homology with mouse chromosome 4, was likewise seen to be capable of suppressing a wide range of human tumor cell phenotypes. Studies on human chromosome 11 found it to be associated with the suppression of malignancy in fusion experiments consisting of human uterine carcinoma and a normal human fibroblast (Harris 1988). Further technical advances enabled researchers to demonstrate directly that single copies of chromosome 11, when delivered by the new method of microcell transfer, were able to suppress the malignant phenotype of either uterine carcinoma or Wilms tumor cells (Harris 1988). Similar methods localized putative tumor suppresser capabilities to nine different chromosomes (Levine 1993). A next step in the logic of this research program was to locate and identify the specific tumor suppressor genes situated on these chromosomes.

Somatic cell hybridization studies were one source of the tumor suppressor gene model; epidemiological studies on the childhood cancer retinoblastoma was another. Retinoblastoma arises from cells of the embryonal neural retina and occurs only in young children. In most cases, retinoblastoma arises sporadically (not along family lines) with an incidence of approximately 1:20,000, yet in about one-third of cases overall the tumor did appear to follow some heritable pattern (Stanbridge 1990). A. J. Knudson in 1971 proposed a "two-hit" model in order to account for this split between heritable and sporadic forms of the cancer. Consistent with the idea of a recessive pattern, Knudson suggested that the same locus was involved in both sporadic and familial forms of the disease but that in the familial form one of the alleles was already mutated in the germ line. A single somatic mutation involving the homologous locus in the unaffected chromosome would then be enough to generate the tumor in the predisposed individual, whereas in sporadic retinoblastoma two mutations must occur somatically in the same retinal precursor cell. Cytogenetic studies corroborated this theory and located the site of the retinoblastoma lesion to chromosome 13, band q14 (Stanbridge 1990).

A third line of research which led in the direction of associating carcinogenesis with the loss of both alleles at some locus was that which sought to correlate tumorigenesis with the loss of chromosomal heterozygosity. In the inherited forms of retinoblastoma, the somatic event that "knocks out" the other allele, thus uncovering the previously recessive germ line mutation, is usually a chromosomal aberration—a chromosome loss, deletion, mitotic recombination, or gene conversion (Marshall 1991). Such events result in the absence of genetic material inherited from one of the parents and can be detected by the loss of heterozygosity for chromosomal markers flanking the locus. Since similar events uncovering recessive somatic mutations occur in the sporadic forms of retinoblastoma, the consistent loss of heterozygosity in tumors can be used as an indication of the presence of a tumor suppresser gene. A putative tumor suppresser gene called p53 was detected on the basis of the correlation of the loss of heterozygosity on the short arm of human chromosome 17 with the occurrence of breast cancer, small cell lung cancer, astrocytomas, and colon cancer (Marshall 1991). This gene has thus far emerged by some margin as the most frequent site at which genetic alterations are associated with human oncogenesis, far surpassing the frequency of correlations found with any of the retrovirally determined proto-oncogenes. In addition to P53 and the retinoblastoma related gene (located on chromosome 13q14), loss of heterozygosity studies indicated the presence of tumor suppressor genes as follows: the Wilms tumor (WT-1) gene located on chromosome 11p13, the adenomatous polyposis (APC) gene located on chromosome 5q21, and the deleted in colorectal carcinoma (DCC) gene located on chromosome 18q21 (Levine 1993).

Cancer Genes: Dominant, Recessive, or None of the Above?

Bishop and Varmus had theorized that the retroviral findings pointed the way to a set of normal growth-related genes, the c-oncs or proto-oncogenes, which become the cause of sporadic cancer when subject to some form of somatic mutation, which became referred to as "activated." A single such event—the activation of any of these proto-oncogenes—would then determine the cancer phenotype and thereby count as a dominant allele. By such reasoning, when a proto-oncogene becomes activated, it acquires a new function by means of which it orchestrates the transformation to a malignant phenotype. In this the activated proto-oncogene resembles the virogene-oncogene of the Huebner-Todero hypothesis except that it is deemed to derive its power from its location within an enzymatic control circuit which regulates cell growth. Cell fusion experiments performed by Harris et al., however, suggested that tumorigenesis can be stopped by the presence of a normal chromosome, implying that in order for carcinogenesis to ensue, it is necessary (regardless what mutational activations may be promoting cancer) that certain (tumor suppressor) genes be entirely lacking. The retinoblastoma and p53 studies appeared to indicate that the lack of certain alleles alone may be sufficient to result in certain kinds of cancer.

In order to support their contention that certain single genes, the proto-oncogenes, could become dominantly acting causes of cancer upon activation, Bishop and Varmus endeavored to demonstrate that proto-oncogenes could be experimentally altered (activated) and made to be capable of inducing neoplastic transformation. The experimental system they used for this work consisted of an aneuploid[2] mouse fibroblast cell line called "NIH 3T3 cells" and a methodology known as transfection for introducing fragments of foreign DNA into these cells. Morphological transformation of the 3T3 cells (which generally meant a change from a flattened to a more spherical shape), loss of contact inhibition (that is, the ability to continue to divide and replicate even after lateral contact with other cells has been made), and an ability to grow in soft agar (which normal fibroblasts are not capable of doing) became the standard criteria for identifying a neoplastic phenotype.

With the recombinant technology for altering DNA sequence and an assay for in vitro transformation it was possible to explore experimentally the meaning of activation.

Much of the experimental work carried out during the 1980s in the quickly expanding field of oncogene research consisted of attempts to clarify the exact nature of activation. Oncogenes and proto-oncogenes from viral and cellular sources were isolated and subjected to manipulation. Various sequences were altered and the effects of the alteration were assayed for carcinogenic capacity through transfection into the NIH 3T3 cell line. The dominantly acting oncogene theory of cancer causation turned on these highly publicized studies. In a scathing 1988 review, Henry Harris took the Bishop-Varmus theory severely to task. His criticisms were enumerated as follows (Harris 1988):

1. It has been shown that the introduction of an oncogene into NIH 3T3 cells or other untransformed cell lines of this type, all of which are aneuploid, produces multiple stable changes in the genome of the recipient cell. Against this complex background of genetic changes, no conclusion concerning the dominance or recessivity of the mode of action of the interpolated oncogene is possible, even when the parameter being studied is no more than morphological transformation *in vitro*.[3]

2. The great majority of morphologically transformed cells are not malignant in the sense that they are capable of progressive growth *in vivo*. When the morphologically transformed cells are injected into an appropriate host, it can easily be shown by karyological analysis that the cells capable of progressive growth *in vivo* (those that generate the tumor) are a highly selected subpopulation.

3. Once malignant cells have been so selected, continued expression of the interpolated oncogene is not required for maintenance of the malignant phenotype.

4. In the few cases where the question has been specifically examined in genuine tumors, it has been found that mutated oncogenes are frequently present in the hemizygous condition.

5. When malignant cells containing known oncogenes that are actively expressed are fused with diploid fibroblasts, malignancy is suppressed whether or not the oncogene remains active in the hybrid cell.

The idea that the NIH 3T3 cell-oncogene transfection system demonstrates that activated proto-oncogenes orchestrate neoplastic transformation in a genetically dominant fashion was attacked by Harris on several levels. The cells that appear to be transformed in culture generally do not prove to be tumorigenic in animals. That only a subset are tumorigenic in animals suggests that something beyond transfection with the oncogene is required. With respect to such cells that are clearly malignant, continued expression of the oncogene does not appear to be required, suggesting that the oncogene is not orchestrating the phenotype. In addition, the aneuploidy of the cell line and the tendency for any transfection to result in numerous genetic changes make it meaningless to speak of genetic dominance because the transformed phenotype cannot be attributed to a single locus, nor can the presence of a wild-type allele be assumed. Finally, even if the genetic characterization of the cells were sufficient to make attributions of genetic dominance meaningful, such an attribution would still be factually inaccurate inasmuch as the transformed phenotype is suppressed in the cell fusion experiments.

An obvious way to have gotten around the genetic ambiguities of the aneuploid cell line would have been to transform ostensibly normal cells. Failure to do so was not based on a lack of effort. As reported by Eric Stanbridge:

Despite intensive efforts to transform normal human fibroblasts or epithelial cells with varying combinations of activated cellular oncogenes, the results have been uniformly negative (Stanbridge 1990).

Perhaps the most convincing evidence against the dominant oncogene thesis came from the then surprising results of high-tech transgenic mouse experiments. Mice were "constructed" such as to contain either the myc or ras oncogenes in every cell of their body. In a characteristic transgenic study an activated myc oncogene had been fused with an MMTV promoter to ensure its expression in the pancreas, lung, brain, salivary gland, and breast of the mouse. Tumors arose only in the murine breast, only on a clonal basis, and only after a latency period (Weinberg 1989). Such results indicated that tumorigenesis had occurred in at most a few out of millions of cells expressing the oncogene, that it was dependent upon additional lesions occurring and that even this effect was limited to cells of a specific tissue and differentiation state. Oncogene researcher Robert Weinberg felt compelled to conclude that

the lesson from this is dramatic and clear: A single oncogene like ras or myc is unable to malignantly transform the great majority of cells in which it is expressed. Although millions of cells in a tissue remain quite normal, only a few will go on to generate a tumor mass. The expression of these single oncogenes may be necessary for tumorigenesis, but is hardly sufficient (Weinberg 1989).

The realization that the expression of a single activated proto-oncogene is not a sufficient basis for cancer should not have been such a surprise when viewed in the light of 50 years of research in experimental carcinogenesis, which had long since ruled against any single-hit model of cancer.

By the end of the 1980s a consensus among molecular oncologists emerged which reendorsed a multistep model of carcinogenesis, postulating a progressive accumulation of lesions to both oncogenes and tumor suppresser genes, with tumor suppresser genes the more apparantly pervasive. The use of Mendelian language, however, did not disappear. Attempting to advance a more genetically ecumenical theory of carcinogenesis, albeit one in which the distinction between oncogenes and tumor suppresser genes was still held to be of more than just historical interest, Bishop redeployed the terms dominant and recessive with an altered specification:

The genetic damage found in cancer cells is of two sorts: dominant, with targets known colloquially as proto-oncogenes; and recessive, with targets known variously as tumor suppresser genes, growth suppresser genes, recessive oncogenes, or anti-oncogenes. The dominant damage typically results in a gain of function, whereas the recessive lesions cause loss of function (Bishop 1991).

Bishop suggested that the distinction between dominant oncogenes and recessive tumor suppressor genes pertained to function. What can this mean? Is this distinction tenable? Where the distinction between the two sets of cancer-implicated genes is least problematic would appear to pertain to biochemical activity. Those genes that have been referred to as oncogenes or activated proto-oncogenes have been implicated in carcinogenesis in relation to the biochemical activity of their gene products. Those genes that have been referred to as tumor suppresser genes have been implicated in carcinogenesis with respect to the loss of expression and/or the absence of the biochemical activity of their gene products. So why not just distinguish these putative cancer-related classes in terms of biochemical activity? The ascription of function is different and stronger than that of biochemical activity. Bishop means to make a stronger claim about the relationship of genes to carcinogenesis than only a distinction in terms of biochemical activity would imply. In analyzing the warrant of his stronger claim we can begin to see the research program in molecular oncology, by its own dynamics, pushing the gene-centered concept to its limits.

The problem with attempting to distinguish oncogenes from tumor suppresser genes in terms of the more empirically tractable criterion of biochemical activity is that the normal c-oncs, or unactivated proto-oncogenes, already have biochemical activity. Where loss of activity could distinguish normal from abnormal states of the tumor suppresser genes, this distinction does not work for the oncogenes. Nor has a change in activity, that is, between the normal activity of the proto-oncogenes and the abnormal activity of the activated proto-oncogenes, ever been established. Indeed, Bishop himself in 1982, and many investigators since, had suggested that there is no qualitative change of activity, only a quantitative increase of the same activity associated with transformation (Bishop 1982). So in the view of Bishop what the proto-oncogene gains in the course of becoming an activated oncogene is not a new biochemical activity but the function of causing cancer. But why speak of the association of the activity, possibly elevated, of some gene product with carcinogenesis in terms of function? Certainly no manner of etiological explanation could justify the use of the term *function* here. Neither the normal proto-oncogene nor its activated state (whatever that might be) is present because it has caused cancer in previous generations.

What is really at stake in ascribing function to the activity of the onco-gene is an attempt to retain a gene-centered interpretation of carcino-genesis. The depiction of cancer derived from the experience of the tumor suppresser research program is a different one. Tumor suppressor genes cannot be the motor-force of cancer causation. The positing of their exis-tence does not constitute a genetically directed story of cancer progres-sion. Where a tumor suppressor function or any other function is lost, the execution of carcinogenesis has not thereby been assimilated into a genetic model. Genetic lesions that result either in the loss of biochemi-cal activity or in an increase of biochemical activity could effect a *loss* of function. As we will see, it is the loss of functions that pertain to ter-minal differentiation that appears to lead the way to malignancy. Again the reflections of Henry Harris will be useful in articulating this alternative framework in which the dominant-recessive distinction is no longer meaningful.

My present ideas on this subject have their origins in a histological observation made by Stanbridge and Ceredig (1981). They found that when cells of a line derived from a human carcinoma were fused with normal human fibroblasts, hybrids in which malignancy was suppressed acquired a different morphology in vivo from segregants in which the malignant phenotype had reappeared as a consequence of chromosome loss. The malignant segregants grew progressively as undifferentiated epithelial tumors; but the hybrid cells in which malignancy was suppressed assumed an increasingly elongated shape and gradually ceased to multiply. Now this is just what normal fibroblasts do in, for example, a healing wound. The cells are at first induced to multiply, but as they secrete and organize their characteristic collagenous extracellular matrix, they gradually elongate and stop multiplying. . . . The observations of Stanbridge and Ceredig thus suggested that the hybrids in which malignancy was suppressed were executing, at least in part, the differentiation programme of a normal fibroblast, whereas the malignant segregants were not.

Essentially the same result was obtained when tumour cells were fused with normal keratinocytes. Terminal differentiation in the keratinocytes involves the synthesis of the protein involucrin, which is cross-linked by a keratincoyte-specific transglutaminase to form an insoluable envelope. . . . When hybrids between normal human keratinocytes and cells of a line derived from a human carcinoma were examined, it was found that those in which malignancy was suppressed continued to synthesize involucrin, but malignant segregants did not. When injected into the animal the non-malignant hybrids showed the characteristic histological features of terminally differentiated keratinocytes and ceased to multiply, whereas the malignant segregants continued to multiply as undifferentiated epithelial tumors. It appears that once again the suppression of the malignancy in these cases involves the imposition on the hybrid cells of the terminal differentiation programme of the normal cell with which the tumour cell is fused (Harris 1990).

In the explanatory model suggested by these findings, cells stop dividing when they enter into a pathway of terminal differentiation. In this view the ability to continue to divide, as is the case in tumorigenesis, does not require the acquisition of new functions but rather is on the order of a default condition of the cell—of any cell—when its progression along a path of differentiation has been impeded. Understood this way, the significance of *either* oncogene lesions resulting in an increase of some biochemical activity or tumor suppresser gene lesions resulting in the loss of some biochemical activity is that in either case it is a disruption of terminal differentiation, that is, a *loss of function*, which is associated with the onset of tumorigenesis. Along with abandoning the idea that oncogenes are characterized by a gain in function goes the specifically gene-centered notion that certain dominantly acting alleles determine the neoplastic phenotype.

Following the idea that some impediment to the pattern of differentiation which is specific to the tissue and stage of differentiation of a cell is the common feature of both oncogene and tumor suppressor gene lesions in the causation of cancer, Harris "invented" a genetic lesion to test his hypothesis. The extracellular matrix molecule fibronectin, which is secreted by fibroblasts, has been shown to provide a mediating link between fibroblasts and other components of the extracellular environment within tissue. Proper attachment of fibroblasts to the extracellular matrix is believed to be a critical step in fibroblast differentiation. Harris transfected nontumorigenic hybrid cells derived from tumor and normal fibroblast fusions with sections of the fibronectin gene in an "anti-sense"[4] configuration. The anticipated result of such a procedure is that of inhibiting the synthesis of fibronectin. The anti-sense fibronectin RNA transcript, which is complementary to the normal transcript, pairs with it to form an inactive RNA duplex, and in so doing it blocks translation and synthesis of the fibronectin protein. Of five clones with fibronectin production abolished or greatly reduced, "in four of these, the malignant phenotype had reappeared and the hybrid cells again produced rapidly progressive tumours when injected into appropriate animals" (Harris 1990).

If one can imagine the fibronectin anti-sense gene arising naturally as a mutation in a cell, one can see the fallacy of the Bishop dominant-recessive gain in function–loss of function distinction. The pattern of involvement in transformation of this gene would follow that of the oncogene and not that of the tumor suppressor gene. Its tumorigenic effects would be associated with its transcriptional activity; the more activity, the stronger the tumorigenic effects. By the Bishop distinction the anti-sense gene would constitute a dominantly acting oncogene with a gain of function. Yet the mode of action of this mutation would be to block the synthesis of fibronectin and thereby impede progress toward terminal fibroblast differentiation. The production of fibronectin anti-sense gene activity in fact represents a loss of function.

The example of the anti-sense fibronectin gene provides a perspicuous occasion for revisiting the Gene-P/Gene-D distinction in the context of carcinogenesis. If we imagine that a fibronectin anti-sense allele[5] came about somehow spontaneously, it would be quite plausible to imagine that its expression could correlate with some elevated incidence of tissue-specific, (fibroblastic) tumor formation at certain developmental stages. Our fibronectin anti-sense gene would then be a proper Gene-P for a certain sarcoma (fibroblastic tumor). If it were in the germ-line, it would indeed follow a Mendelian pattern of inheritance. As a Gene-P (which could be identified with a molecular probe), it could do what a Gene-P *can* do. It could pragmatically serve as a predictor of an elevated cancer risk at certain times and under certain conditions. Where conflationary confusion arose (and this is exactly what Bishop's analysis entails), it would be treated as a specific molecule (Gene-D) which orchestrates (programs, dictates, determines, controls, effects, and so forth) the tumorigenic phenotype. The capacity and proclivity for fibroblasts to undergo abnormal growth in the absence of secreted fibronectin—which would be restricted to certain tissue-developmental and microenvironmental contexts—would not be explained, let alone programmed (dictated, executed), by the absence of the fibronectin. An unconflated Gene-P (that did not confuse prediction with function) would not masquerade as a substitute for understanding the actual biology of neoplastic growth.

Despite extensive efforts, no mechanistic pathway has been identified to account for the means by which activated proto-oncogene products function to transform cells. Most likely, as in the case of the anti-sense fibronectin gene (but with less specificity), the activated proto-oncogene products act to disturb processes of differentiation in certain cells at certain times, resulting in a loss of the ability of these cells to obtain or retain a state of terminal differentiation.

Further examination of the nature of cancer-related genes lends support to Harris's theory. Consider the case of colon carcinoma, which has been used as the basis of a more recent multistep genetic model of carcinogenesis (Fearon & Vogelstein 1990). The basic schema proposed for tumor progression in the colon is one which begins with lesions to the gene APC followed by an activating lesion to the oncogene ras and then sequential loss of the DCC and p53 tumor suppressor genes. While ostensibly another attempt to make good on the "genetic orchestration of cancer" concept, a closer examination of the evidence suggests the loss and/or disruption of stabilizing functions and a permissive rather than determinative model of the role of mutations.

One in 10,000 individuals in the United States, Europe, and Japan suffers from an autosomal-dominant disease known as *familial adenomatous polyposis* (FAP). Individuals with this affliction develop thousands of benign adenomatous polyps of the colon during the second and third decade of life, with a small percentage of these becoming carcinomas of the colon. The gene APC, located at chromosome 5q21, was identified as the site of the relevant lesion (Levine 1993). Recent studies have indicated that the APC gene product appears to participate in the protein complexes associated with cell-to-cell binding and tissue-stabilizing junction formation (Hulsken, Behrens et al. 1994). Along similar lines the DCC (deleted in colorectal cancer) gene which is found to be lost in more than 70 percent of colorectal cancers has a gene product with significant amino acid homology to that of the neural cell adhesion molecule (NCAM) (Levine 1993). Reminiscent of Harris's constructed fibronectin mutation, the identification of APC and DCC as cell-cell interaction related genes suggests that such interaction is a requirement for successful colon cell differentiation and that these genetic lesions represent a loss of this function.

According to the colon carcinoma model, homozygous deletion of the tumor suppressor gene p53 occurs at the latter stages of the carcinogenic sequence. Thus far, p53 has been implicated in a greater array of human cancers than any other gene. Like Rb (the retinoblastoma tumor suppresser gene), and the oncogene myc, the p53 gene product binds to DNA and appears to be an effector of transcriptional activity (Haffner and Oren 1995). The effects of p53—like those of myc, Rb, and probably all transciptional regulators—are highly complex, being capable of either up-regulating or down-regulating transcriptional activity, depending on the physiological-biochemical context. The gene product of myc, for example, may serve either in the stimulation of cell proliferation or in apoptosis (described as programmed cell death), depending on the prevailing state of cytokines[6] (Wyllie 1995). Increasing amounts of attention have been directed toward evidence of cancer-relevant roles of p53 in cellular responses to genetic damage. Two different pathways, one toward arrested growth and the other toward apoptosis, appear to be concerted responses to genotoxic challenge, in which the p53 protein plays a role (Haffner & Oren 1995).

Following Harris, we have explored the possibility that the most reliable correlation of genetic damage with cancer can be best interpreted as a loss of some developmental resource (Gene-D) which had become necessary for the continuation (or retention) of development *specific to a particular path of differentiation*. The identification of important tumor suppresser genes as putative cell adhesion molecules corroborates this view. The p53 gene, which thus far has shown the widest range of applicability to human cancers, may do so because it serves as a necessary developmental resource in a kind of meta-differentiation which transcends tissue-specific differences. Apoptosis may be best conceptualized as a meta-level form of terminal differentiation that can be induced by genetic damage and other events. In this model of p53 action, as well as that of APC and DCC, the significance of genetic damage is that of a *permissive* and in *no manner directive* relationship to carcinogenesis. The lesson to be learned from Harris and the tumor suppressor research program is that the question of the driving force of carcinogenesis either cannot be answered at the level of genetic analysis or perhaps is just not the right question at all.

Life at the Margins: Adapting to Altered Circumstances

I have argued that a bifurcation took place in thinking about cancer early in the 20th century with the main trend being toward what I have referred to as "the phylogenetic turn" and also as gene-centrism. The legacy of the nineteenth century developmentalist outlook on cancer was that of grappling with the consequences that would follow from the fact of any cell's being invested with the *Keime und Anlagen* of the whole organism. If cells retain the potential of the whole, then the fate of a given cell cannot be determined solely from within but must be largely the consequence of subsequent interactions with other cells and acellular elements of its surround (including environmental irritants). I have also suggested that the gene concept that arose in the twentieth century had roots in a need to chop up the *Keime und Anlagen* in order to square it with the apparent requirements of Darwinian evolutionary theory. The somatic mutation hypothesis put forward by Boveri brought this turn of mind to bear on cancer research and established a conception of the cancer cell as determined from within. From these two legacies arose two distinctly different conceptions of the autonomous cell, as discussed above. Sections 2 through 4 largely followed the fortunes of the somatic cell hypothesis with its equation of cellular autonomy and the so-called cancer cell. The somatic mutation hypothesis reached its apex with the Bishop-Varmus oncogene model but ultimately failed to support its most interesting claims against challenges, both implicit and explicit, that arose from parallel studies on the role of recessive genes in carcinogensis. Section 5 "thematized" the nature of these challenges.

The subject of cancer raises with poignancy this question: At what level of biological order can the distinction between normal and pathological be properly made? The prevalent focus on the nature of the cancer cell has been misleading to the extent that it has masked a more divergent and interesting conflict over the locus of normality-abnormality. While somatic mutation theories have attempted to implicate genes as being the right level for this analysis—that is, the subcellular level—the older idea that it must be at the *super*cellular level that normal versus pathological growth trajectories are determined, was never completely vanquished.

A particularly vociferous attack on the reductivism of the cancer cell theory was launched in 1962 by the British radiologist D. W. Smithers in the prestigious medical journal *The Lancet*. As the epigraph at the head of this chapter suggests, Smithers presented what could well be described as a manifesto calling for not just a new cancer biology but also a new biology of the life process and organization without which, he would claim, cancer could never be adequately comprehended. His style of attack was eminently straightforward. He enumerated the components of the mainstream reductionist view, then laid out the incompatibilities between it and the accumulated experience of his own career in pathology and clinical radiology. He then adumbrated the elements of his alternative model. Despite the passage of over 35 years, his rendition of the mainstream view still sounds strikingly familiar, his anomalies are still anomalies, and his alternative model is still alternative and still interesting. I will present all three.

Smithers's characterization of the traditional view of cancer (by "traditional" Smithers really means the mainstream view of his day and certainly not anything prior to Boveri) is as follows (Smithers 1962):

1. Cancer is a special disease of cells.

2. A cancer cell is one that has been permanently changed and is no longer capable of behaving like a normal cell.

3. Cancer cells multiply without restraint and produce tumours serving no useful purpose in the body.

4. Cancer cells grow at the expense of normal tissues, actively invading and destroying them.

5. Cancer cells can gain access to cavities and to lymphatics and blood vessels and be carried off, each one being capable of developing into a new tumour wherever it may come to rest.

6. If the cause of cancer in the cancer cell could be discovered, the whole problem might be resolved.

7. Cancer cells must be removed or destroyed in situ if patients are to be cured.

8. If one viable cancer cell remains, treatment will fail.

9. A chemical poison specific to the cancer cell may one day be found to replace all present treatment methods.

10. When disease has spread beyond the scope of local removal or irradiation, any nonspecific cell poison may be worth trying until something better comes along.

This list provides a strikingly apt representation of the suppositions that have guided both basic and clinical research in oncology over the last 40 years. Item number 6—the idea that if we could ascertain the determination of the cancer phenotype within the cell we would solve the problem of cancer—has been the guiding motivation behind all the research in molecular oncology. In this respect, both the oncogene and the tumor suppresser gene concepts partition squarely on the far side of

Smithers's divide. Smithers's analysis well describes all the implications of the autonomous-cell-as-cancer-cell equation and in so doing highlights how many taken-for-granted "facts" about cancer would have to be made problematic if the basic notions expressed in, for example, numbers 1 and 6 were to be reconsidered.

Smithers's list of incompatibilities is as follows:

1. The multicentric origin of neoplasia. (Many cancers appear to be derived from more than one cell.)

2. The long prediagnostic natural history and the many predisposing factors in the development of tumours. (Histologists never see a radical transition in cancer, only gradual changes along a continuum over time.)

3. Age incidence and geographical variation.

4. Progression and regression in tumor behaviour. (The experience of spontaneious regression is of particular interest in this regard.)

5. The conditional persistence of some tumors. (Some tumors appear to continue to be dependent on certain environmental conditions.)

6. The hormonal dependence of other tumors. (The status of a cell as a cancer cell may be hormone-dependent.)

Smithers's list consists of those frequently observed aspects of carcinogenesis that point away from a strictly internal (to-the-cell) basis of cancer and toward the relevance of higher levels of organization. None of these six points has become irrrelevant during the ensuing 40 years; rather, attempts have been made to accommodate such observations with somatic mutation hypotheses such that they do not appear to be wholly anomalous. But this would be the pattern seen for any number of research programs that begin to run out of explanatory steam. Even if alleged contradictions can be eased or softened, there is still the danger of each new emendation to the model increasingly taking on an ad hoc character. As an alternative to piecemeal attempts to fix the somatic mutation theory Smithers offered what amounts to the reassertion of a developmental-organizational perspective as follows:

1. Cancer is a disease of organization.

2. The word "cancer," however, merely covers the most disorganized end of a progression in disorganization extending from maldevelopment,

malformation, metaplasia, hyperplasia, dedifferentiation, and neoplasia to disintegration.

3. There is no such thing as a cancer cell—only cells behaving in a manner arbitrarily defined as being cancerous.

4. There being no such thing as a cancer cell, there can be no cause of cancer to be found within it.

5. Organization is measured by the amount of information in a system, and the entropy of a system is a measure of its degree of disorganization; gradual increases in entropy occur, not sudden cellular malignancy.

6. Organizational breakdown commonly leads to progressive loss of growth control, with released cell division producing an excess of tissues no longer coordinated with the whole.

7. Organization can become more or less disorganized, and the resulting tumours may progress or regress in behaviour patterns.

8. An abnormal cell, particularly a stem cell, may produce a clone of cells reacting abnormally with the environment and so promote disorganization.

9. Environmental stress may disorganize the behaviour of many cells within a sphere of influence which may be local or widespread, depending on the character and distribution of the particular cells concerned.

10. There are many causes of organismal disorganization; understanding depends on the explanation of tissue dependencies and relationships which are complex, changing, and never likely to be totally foreseeable.

At first blush, Smithers's view would appear to be far outside the current consensus, holding as it does that cancer is unmistakably the result of genetic lesions; many would be inclined to dismiss it simply for being hopelessly outdated. Yet, with critical examination we can see that this is misguided. Smithers suggests that carcinogenesis is a gradual process, whereas the somatic mutation viewpoint has traditionally focused on the idea of oncogenic mutation as an all-or-nothing event. Current thinking however, has once again rejected the idea that cancer can be caused by a single somatic mutation. The leading exemplar of a multistep somatic mutation model of cancer, colon carcinoma, is understood to require four to six, or even more, sequential mutations. Clearly, this must constitute a *gradual* process. One may well try to argue that while sequential and gradual, the colon model still describes an entirely *intra*cellular process. Yet even this can no longer be considered a secure presumption. The (cellular) genome is quickly emerging as a far more dynamic system than had even recently been imagined. Genetic rearrangement, on many levels, is mediated by active cellular processes. Once it becomes established that mutations are not exclusively, or perhaps even principally, the result of biologically blind physiochemical chance (e.g., quantum fluctuations in electron distribution) but rather the result of biological processes such as enzyme systems, then the sharp intracellular versus extracellular distinction becomes untenable because the processes that are mediating genetic rearrangements are themselves susceptible to the influences of the larger extracellular field. Even mutations that are the result of a general downregulation in mutation repair systems can be linked to the effects of the larger extracellular field upon the mutation-repair state of the cell. By "extracellular field" I am referring to the sum total of cell-cell, cell-matrix, receptor-ligand, chemical messenger, local ionic concentrations, and so forth, which influence all the activities of a cell.

Perhaps a particularly compelling illustration of this point is the emergence of gene amplification as a major type of somatically based genetic rearrangement found to be associated with human cancers. *Amplification* refers to the actual multiplication of the number of gene copies within the DNA of a cell. It turns out that amplification of proto-oncogenes is one of the major pathways by which the proto-oncogene is activated. Amplification is found to result in from 5 to more than 500 copies of the original gene becoming present and does not typically entail a change in DNA sequence, i.e., a classical mutation (Schwab 1998). The activation of proto-oncogenes by amplification is a matter of gene dosage. The oncogene ERBB2, for example (also known as HER2 and Neu), is found to be amplified in 20 to 25 percent of primary breast cancers although no sequence mutation is associated with it (Schwab 1998). While all the mechanisms associated with gene amplification are not known, there can be little doubt that it involves an array of complex biological activities. Enzymes, in all likelihood, mediate the polymerization of new DNA using the old gene as a template, and additional enzyme systems must make the necessary incisions and ligations needed to result in a continuous strand of de novo elongated DNA.

At this level of biological complexity the idea that such events take place at random in hermetic isolation from all the ambient factors that influence every other biological process is simply untenable. The emerging consensus of a multistage multimutational basis of cancer makes reference to an increasingly complex shopping list of factors, oncogenes, tumor suppresser genes, cell-adhesion molecules, and so forth that are essential ingredients of malignancy. But where is the line between cause and effect? On what basis can it be held that subsequent alterations of cellular DNA, amplifications, rearrangements, and the like, which are themselves mediated by biological subsystems, are not the effects of biological factors, including the larger extracellular field, every bit as much as they are the effectors of subsequent biological events?

Smithers may not have been privy to the wide array of genetic events that have since been shown to be associated with carcinogenesis, but it is the very nature of these genetic events that is beginning to point the way back to the significance of the larger extracellular field. Particularly striking in this regard is the emergence of members of every class of cell-adhesion molecule (those cell surface components which are most directly involved in mediating cell-cell and cell-matrix interactions), i.e., the integrins, cadherins, and the immunoglobin superfamily, in cancer-related alterations. Changes in the composition or character of cell-adhesion molecules may directly affect both the influence of a cell on its extracellular surround and how it in turn is affected by that surround. That somatic alterations of cell-adhesion molecule genes are emerging as frequent factors in carcinogensis would strongly suggest that the internal milieu of a cell can't be separated from its larger context when considering the etiology of cancer.

The somatic rearrangement of various genes in carcinogenesis has been well established and yet neither the onset of these rearrangements nor the consequences of these rearrangements occurs in isolation from the larger extracellular milieu. Smithers's idea that cancer is principally a disease of extracellular organization is thus not overruled by the last 40 years of findings in cancer genetics. Nor has thinking along these lines gone entirely unnoticed by the genetic mainstream. Consider, for example, the following early reflection from leading oncogene investigator and Nobel Laureate J. Michael Bishop:

Each oncogene induces tumors in only a limited and characteristic set of tissues: transformation of cells in culture follows the same selective pattern. We cannot at present explain the selectivity of oncogene actions, but the phenomenon has contributed to the view that transformation by retrovirus oncogenes is fundamentally a disturbance of differentiation. According to one prevalent view, oncogenes may act by arresting cellular development within a specific compartment of one or another developmental lineage; tumorigenesis ensues because the immature cells that constitute the compartment continue to divide, as is their nature, and become a continuously expanding population (Bishop 1982).

Smithers (point no. 8) allowed for the possibility of an abnormal cell which results in a clone of cells that react aberrantly with their environment. The anonymous, yet prevalent, view to which Bishop refers would appear to be akin to a kind of deflationary account of the role of oncogenes in cancer. Oncogene activation would be seen not as the determinant of cancer but as the proximate initiator of some kind of abnormality, the consequences of which are contingently determined by the relevant context. Bishop did not choose to turn down that interpretive path, and the mainstream of cellular cancer research has continued to focus on the strictly intracellular, genetic determination of the cancer phenotype. Yet the evidence for the context specificity of cancer has not declined in the slightest; nor, given the emerging knowledge of the biologically mediated nature of somatic genome alterations, can Smithers's idea that carcinogensis may begin with tissue disorganization be ruled out. If Smithers's (1962) view appears ostensibly to undervalue or ignore the significance of intracellular events, it would seem that molecular oncologists up to the present have persisted in dogmatically adhering to a gene-versus-all-else dichotomy, when precisely it is the need to understand the multidirectional vectors of causality that is indicated. A stunning illustration of the inextricability of carcinogenesis from the larger developmental field can be found in the biology of those vertebrates that have retained substantial regenerative capability.

Substantial somatic regenerative ability is restricted in vertebrates to members of the *Urodele* amphibians, which includes newts and others salamanders. An adult newt can regenerate its tail and limbs as well as upper and lower jaw and ocular tissues such as the lens. This process, which is called "epimorphic regeneration," begins with the formation of a local growth zone called a "blastema." *Urodeles* can reverse the differentiated state of tissues in response to amputation or tissue removal (Brockes 1998). De novo undifferentiated cells proceed to undergo several rounds of division before redifferentiating into a new lens or limb mesenchyme.

Urodeles in general, and especially the particular *Urodele* tissue that is possessed of regenerative capacity, show striking resistance to tumor formation. Evidence suggests that in response to experimental exposure to carcinogens, supernumerary regenerates form. It is conventionally held, in concert with the somatic mutation model of carcinogenesis, that cell senescence—the apparent limitation on the number of times a cell is capable of dividing—is a built-in safeguard against cancer, resulting from the accumulation of multiple mutations. Yet the tissues of the *Urodeles*, which are capable of forming a blastema and thus have indefinite growth potential, are the vertebrate tissues that are most resistant to tumor formation. There is no basis for imagining that these tissues are any less susceptible to somatic mutations, indeed only more so owing to their growth capacity. It is thus suggestive that tumor resistance is a function of the ability of the multicellular milieu to enter into a de novo morphogenetic pathway. Brockes (1998) described the cellular response of *Urodele* tissue to "oncogenic activation or loss of tumor suppressor function" as the return of differentiated cells to the cell cycle, dedifferentiation and then participation in regeneration-type patterns of behavior rather than tumor formation. "It is possible . . . that if tumorigenic mutations arise, the cells are somehow constrained within the regulatory framework of epimorphic regeneration" (Brockes 1998).

With additional experimental data, the *Urodele* example could well provide powerful support for a renewed version of Smithers's thesis. It appears likely that no number of oncogenetic mutations can turn a blastemal cell into a cancer cell because cancer is simply not determined at a single cell level, and the higher-order structure of blastema-forming tissues is in effect poised to elicit the behavior of newly proliferative cells (whatever route, mutational or otherwise, they followed to newly proliferative status) such as to result in de novo organizational integration.

More recent consideration of the kind of kinetics and possible mechanisms associated with chemical carcinogensis lends even further support to Smithers's view. Experimentally proven carcinogens are actually highly variable with respect to their mutagenicity. But regardless of whether a chemical carcinogen is shown to be mutagenic or not, there are a number or characteristics of chemical carcinogens that are virtually invariant. Chemical carcinogens are not promoters of cell growth but rather inhibitors of cell growth. Cancer development, as demonstrated with the use of chemical carcinogens (as well as radiation, DNA viruses, and some RNA viruses), is a prolonged process requiring from one-third to two-thirds of the life span of an organism (Farber, 1991). Initiation (of the process of carcinogenesis) with a carcinogen is never immediately followed by spontaneous or autonomous proliferation of cells of any organ. The unrestrained growth characteristic of cancer is a property of the latter stages of the process of tumorigenesis. The kinetics of autonomous growth as seen in experimental systems using chemical carcinogens would thus support Smithers's idea that cancer is not the result of a rapid change within the cell that causes it to become autonomous. What then causes the onset of chemically induced carcinogenesis if (1) the mutagenicity of a chemical carcinogen is not relevant to its carcinogenicity, (2) chemical carcinogens do not promote growth but rather inhibit it, and (3) the acquisition of unrestrained growth occurs only late during the course of a very lengthy process? Could the biological significance of carcinogens be not that of inducing uncontrolled growth through bringing about somatic mutation but rather that of serving as an organizational irritant or disrupter which triggers adaptive but destabilizing responses in cells and tissues?

A possible clue was derived from the cell culture studies of Mondal and Heidelberger (1970). By exposing cells derived from mouse prostate to the potent carcinogen methylcholanthrene, they found that *all* exposed cells gave rise to clonal populations out of which some minority of cells gave rise to transformed foci. Studies showed that some alteration had taken place in 100 percent of the exposed cells. Each and every exposed cell had become capable of giving rise to progeny cells, out of which a smaller subset then produced transformed (tumorigenic) colonies. Of nontreated control cells, by contrast, only 6 percent would ultimately give rise to a colony producing progeny cells. Population-wide (100 percent) responses do not fit the profile of a mutation (which is always a low probability event) but rather are suggestive of a physiological (epigenetic) phenomenon (although a general tendency toward diffuse, nonspecific genetic damage might also fill the bill). Observations, such as those of Smithers, that carcinogenesis often begins from a whole (multicentric) field of cells, have been made in parallel with findings that have supported the theory of the monoclonal (from one cell) origins of cancer. How best to reconcile this apparent contradiction has been given little attention; rather, evidence of the monoclonality of cancers has just been taken as confirmation of the somatic mutation model and thus of the purely internally determined cancer cell, leaving the question of field effects largely ignored. Yet there are many possible explanations for how cancer could begin at the level of a disrupted field and lead to a monoclonal tumor, which would highlight rather than dismiss the importance of cellular interactions within an organizational field.

The principal difference between an organizational field model for explaining carcinogenesis versus a somatic mutation model is that of the hierarchical level which is being examined. An organizational field story cannot and should not attempt to exclude intracellular events, and intracellular events would certainly include changes in the structure and activity of DNA and chromosomes. What an organizational field approach would eschew would be the attempt to treat an intracellular event as a self-sufficient determination of a carcinogenic trajectory, which is exactly what the somatic mutation tradition has attempted to put forward. So the question for an organizational field analysis is not whether genetic alterations occur in carcinogenesis—they surely do—but how to situate them in the complex nexus of causes and effects. Tumor progression entails increasing genetic instability. Genetic instability and its consequences are in turn largely mediated by the organizational context of the cell.

The age of a tissue (and its developmental status) has long been seen as highly significant with respect to both the likelihood of giving rise to a tumor cell and with respect to its receptivity to tumor growth. In classic early studies, Mintz et al. (1978) produced a mouse teratocarcinoma by transplanting 6-day-old mouse embryos to under the testis capsule of an adult male mouse. If these cells were then subsequently injected under the skin of a mature mouse, they were seen to form tumors. Yet if the same cells were inoculated into a very early embryo they became integrated into the developmental matrix of the embryo, becoming normal constituents of many different tissues.

In a conceptually similar and more recent set of experiments (McCullough et al. 1997), liver cancer cells from a rat were transplanted into the livers of both older and younger mice. In the older mice the cancer cells were highly likely to produce a tumor. In the younger mice the same cells were prone to differentiate into normal liver cells. There is striking evidence that the tissue context in which many human cancers develop is different from that of normal tissue. A gradient of biochemical and cytological abnormality has been observed extending some distance from the edge of bladder tumors (Rao et al. 1993). The abnormality of adjacent tissue may include genetic features, or it may not. Ostensibly normal breast tissue adjacent to breast tumors shows loss of alleles (Deng et al. 1996). Victims of Barrett's esophageal cancer have chromosomal changes in large areas of the esophagus in which cancer later arises. By contrast, mucous tissue around colorectal cancers displays various phenotypic abnormalities but no genetic changes (Boland et al. 1995).

The incidence of human cancer is highly correlated with increasing age. In an attempt to sort out the heterogeneity of age correlations that does exist, human cancers were classified into two groups (Dix & Cohen 1980). Class 1, which comprises the great majority of human cancer (including tumors of the mouth, esophagus, stomach, colon, rectum, liver, pancreas, bladder, brain, bronchus, trachea, myeloid and other non-lymphatic leukemias, prostate, and penis), shows a smooth logarithmic increase in incidence between ages 10 and 80 years. The second class, which includes lymphatic leukemia, bone, testes, and Hodgkin's cancers, shows two peaks of incidence, one around the age of 35 and the other at greater than 50 years. The most pronounced correlation would be that of prostate cancer (class 1); 80 percent of American men are found to have clinical prostate cancer by the age of 80.

Experimental studies on tumor-age correlations support the epidemiological findings. When the environmental carcinogen N-methyl-N'-nitrosourea was inoculated into mice at 3, 12, and 24 months of age, only the oldest group developed pancreatic cancer (Zimmerman et al. 1982). Pancreatic cancer is extremely rare in humans below the age of 40. Age-based differences in susceptibility to cancer can even be seen in cells in culture. Cells from the bladder epithelium of old mice are far more susceptible to chemical carcinogenesis than those from young mice (Summerhayes & Franks 1979).

While aging may result in a gradual loss of tissue durability and plasticity and thus the ability to assimilate perturbations, loss of the organizational stability of tissues may also be the result of adaptive responses to acute challenges. Liver biology provides a purchase upon organizational dynamics that appears to be instructive on a number of levels. Liver cell cancer affects hundreds of millions of people, ranking second worldwide, although unlike most of the "top ten" cancers it is less common in the Western world. There is increasing evidence that in parts of China and Africa it is correlated with exposure to mycotoxins (toxins produced by fungi) such as aflotoxin, and in Western countries there is a similar but less pronounced correlation with alcohol consumption and cirrhosis of the liver (Farber 1987). The mechanisms by which etiological agents are involved in the causation of liver cancer are unknown, but what does appears to be uniformly the case—whether it be mycotoxins, alcohol, or hepatitis B virus—is that hepatocellular carcinoma (liver cancer) is a long process, requiring 30 to 60 years from the time of initial exposure. Liver cancer is evidently a multistep process, and attempts have been made, beginning with Sasaki and Yoshida (1935), to establish an experimental model by means of which the sequential stages of liver cancer can be characterized.

Beginning in 1976 Farber and coworkers identified the initiation of liver cancer with the formation of a "resistant hepatocyte." A successful experimental model for the study of cancer development requires an ability to identify the initiation of cancer through some form of assay and a sufficient degree of synchrony in the subsequent progression of initiated cells such that discrete stages can be identified and characterized (Farber, 1987). The rat liver afforded Farber and coworkers just such an opportunity. With exposure to over 75 different chemical carcinogens a small number of liver cells were seen to take on a new phenotype described as "resistant." This resistant phenotype could be characterized in three ways: (1) Resistant cells can be induced to grow, whereas the majority of hepatocytes are growth-inhibited by the carcinogen; (2) Resistant cells exhibit highly enhanced tolerance for cytotoxins; (3) Resistant cells have a distinctive profile of enzymes consistent with their ability to withstand cytotoxic challenge (Farber 1991). All three of these features are suitable for assay.

Farber defined the process of initiation as follows:

Initiation is a change in a target tissue or organ, induced by exposure to a carcinogen, that can be promoted or selected to develop focal proliferations, one or more of which can act as sites of origin for the ultimate development of malignant neoplasia (Farber 1987).

It should be noted that the first stage in Farber's model of carcinogenesis need not itself be an *aberrant* response to a challenge. Indeed, the resistant phenotype has the earmarks of a physiologically *adaptive* response to hepatotoxic insult. What signifies the resistant phenotype as a stage in the process of carcinogenesis is not its morbidity but rather its relationship to other stages in the process. The subsequent stages are not inevitable. As has been seen since the earliest work on experimental models of carcinogenesis, initiated tissue only proceeds towards malignancy under conditions of promotion.

Promotion is the process whereby an initiated tissue or organ develops focal proliferations (such as nodules, papillomas, polyps, etc.), one or more of which may act as precursors for subsequent steps in the carcinogenic process (Farber 1987).

Promotion may be accomplished in the rat liver model by means of renewed exposure to cytotoxins, dietary deficiencies, or some other regime of protracted metabolic stress. Depending upon the nature of the regime, microscopic foci or islands of resistant cells may predominate in the liver for weeks or even months or may rapidly give rise to visible nodules. Several lines of evidence suggests that nodule formation constitutes a highly organized developmental process which serves a physiologically adaptive function in protecting the organism from exposure to toxins. Chief among such evidence is exactly the newly acquired ability of the organism to withstand high doses of hepatotoxins.

Rats with hepatocyte nodules are unusually resistant to a lethal dose of a potent hepatoxic agent, carbon tetrachloride. Rats with nodules show a complete resistance to a dose of carbon tetrachloride that is lethal for 100% of normal rats (Farber 1991).

Further support for the physiologically adaptive nature of the hepatocyte nodules is found in the ability of the majority of the nodules to undergo a complex process of redifferentiation, becoming normal adult-like hepatocytes. That the majority of nodule hepatocytes are capable of returning to a normal phenotype provides strong evidence that the formation of the resistant phenotype, with its ability to proliferate into nodules, is a process that lies within the normal developmental prerogatives of the liver cell. And yet the diminished vulnerability to hepatotoxins that the liver nodule affords the organism may be purchased at a price. In Farber's rat liver model, 2 to 5 percent of the nodules do not redifferentiate into normal adult liver tissue but rather persist. It appears to be at this point that the phenotype of the nodular hepatocytes is no longer a response to ambient conditions but begins to represent an uncoupling from its microenvironment and a shearing off to an independent developmental trajectory.

The persistent nodules acquire a new property—"spontaneous" or seemingly "autonomous" cell proliferation of their hepatocytes—and become the origin for a slow evolution to hepatocellular carcinoma (Farber 1987).

The path from persistent nodules to malignancy is characterized by a stagewise progression, albeit one that eventually becomes difficult to categorize. Two months after initiation, a 10-fold greater propensity for growth among hepatocytes of persistent nodules is observed relative to that of the surrounding liver, but growth is still largely synchronous and well organized. By 6 months the baseline percentage of proliferating hepatocytes was seen to double. Between 2 and 6 months changes in the growth potential of the nodule hapatocytes was observed. All hepatocytes at 2 months (control, nodule, or nodule surrounding) can be stimulated to grow by use of mitogens or partial hepatectomy, but all three then return to basal growth rates soon thereafter. However, similar treatment of 6-month nodules resulted in a proliferative response on the part of 60 to 80 percent of the hepatocytes, with a significant number failing to return to the baseline rate for several weeks thereafter (Farber, 1987).

The persistent nodules at 6 months also show another new property—generation of nodules and hepatocellular carcinoma on transplantation to the spleen.

The earlier nodules, like normal live hepatocytes, grow slowly in the spleen with gradual replacement of the splenic pulp but without nodules or cancer (Farber 1987).

The legendary baseball player Mickey Mantle, publicly famous for home runs and enduring pain, and privately distinguished for sustained alcohol consumption, became fond of saying that if he knew he was going to live so long, he would have taken better care of himself.[7] Mantle died in his early sixties of liver cancer soon after having received a somewhat controversial liver transplant. Mantle's liver cancer may well have been clonal, but it was hardly the outcome of some chancy somatic mutation occurring in isolation from the larger organizational field of his liver. Mantle's liver underwent rounds of organizational shifts in the course of decades of heavy alcohol exposure, which enabled him to enjoy the longevity sufficient for a Hall of Fame baseball career and a subsequent (albeit less distinguished) run as a Manhattan restaurateur. The liver cancer, clonal or otherwise, was the ultimate result of cellular dynamics that were highly adaptive but came at a price.

The apparent paradox of the "multicentric" origins of cancer (to use Smithers's term) and the ultimate appearance of a monoclonal tumor appear to be reconciled in the cell culture model of carcinogenesis developed by Harry Rubin. Four decades after establishing the experimental basis for studying the retroviruses, Rubin continued to be engaged in an experimental practice that he has described as his "dialogue with cultured cells" (personal communication). Rubin's experimental system has involved the use of the same cells, the NIH 3T3 line derived from mouse fibroblasts, that oncogene researchers used for showing the transforming capability of viral and activated oncogenes. But Rubin found that the very same indications of spontaneous transformation in culture are routinely obtained on the basis of metabolic stress, especially the growth of cells under crowded postconfluent conditions. Rubin has found, over years of exhaustive trials, that protracted exposure of cells to conditions of confluence (when the entire floor of a tissue culture dish is filled with a complete pavement of cells one layer thick) results first in both the death of some cells and a population wide change in most. These cells, when switched to optimal growth conditions, reveal a decreased growth rate, yet a greater saturation density, that is, an ability to continue to grow beyond the typical contact constraints of (*so-called*) normal 3T3 cells (Rubin et al., 1995a). Successive rounds of prolonged incubation of these cells at confluence results in the appearance of transformed foci, dense colonies of cells derived from a single cell which are no longer constrained to grow in monolayer but rather form a kind of multilayered "cell culture tumor." Whereas the early changes in the 3T3 cells appear to occur on something approaching a population-wide basis, it is only a small percentage of the cells that are responsible for forming transformed foci. The early changes that take place on a widespread basis, which result in reduced rates of cell growth but higher rates of saturation density, have the character of adaptive epigenetic changes. And these cells do not immediately revert to normal growth patterns when transplanted to low-density cultures (Rubin et al., 1995b). A population-wide response is not consistent with the expectations of a *specific* mutation, which would be a very low probability event. Yet this observed heritability would be consistent with some form of genetic alteration. The early changes of cells grown in postconfluent cultures have characteristics of both epigenetic adaptation to stressful conditions and some form of irreversible genetic alteration, but these need not be contradictory. Adaptive epigenetic changes can also result in genetic instabilities. A possible mechanistic pathway could entail changes in DNA methylation states that can affect mutation rates (discussed in Chapter 3). While the widespread character of the changes seen in postconfluent cultures rules out specific point mutations, diffuse genetic damage need not be excluded. In fact these stages, beginning with a general decrease in growth rate and culminating in the formation of sporadic transformed foci, bear a strong resemblance to in vitro studies in which cells were initially irradiated with X rays or exposed to various chemical carcinogens (Rubin et al., 1995b). What appears to be the common denominator is that cells confronted with any number of stresses respond in an arguably adaptive fashion that allows the majority of cells to continue to maintain a relatively normal phenotype but with an increasing propensity to give rise to a tumorigenic colony. Rubin et al. (1995b) suggest that "All the evidence points to the origin of cancer from a field of altered, unstable but normal-appearing cells rather than from isolated mutations among otherwise unaltered cells."

When Rubin's cells are subjected to further rounds of incubation under conditions of confluence, they eventually give rise to some array of transformed foci which are each in some sense morphologically distinct. A given dish will thus have several different transformed populations with distinctive phenotypes at the colony level. Rubin found that if the cells of these dishes were to be further passaged (removed and replated onto new dishes at initially lower densities), the cultures would eventually become monoclonal cultures of transformed cells. What this suggests is that the ultimate appearance of a monoclonal tumor population in Rubin's experimental system is the result of the fastest growing transformed population, overwhelming the other cells over time. Rubin's experimental system thus provides a model for how cancer may begin as a widespread multicentric disruption of a whole cellular population (and organizational field) yet result in a monoclonal tumor. If the eventuality of a monoclonally derived tumor need no longer appear contradictory to the multicentric origins of carcinogenesis in the disruption of an organizational field, then the most interesting question may pertain to the microinteractions within the disrupted field that result in the promotion and amplication of a clonal tumor. Rubin (1993) has described this process as that of a "progressive state selection." "Our current work goes beyond the monoclonal origin of cancer (Fialkow) to the fields surrounding the dominant clone to establish a cellular ecology of tumor development" (personal communication).

Under the rubric of "dynamic reciprocity" Mina Bissell (a former postdoctoral student of Rubin) and coworkers have been attempting to elucidate the molecular details of the cellular ecology of breast cancer development for more than 20 years. Given the well-known inductive significance of tissue interaction in the case of embryological development, Bissell has approached breast cancer with an eye to the reciprocally stabilizing and destabilizing interactions between the glandular epithelium of the breast and the fibroblastic cells of the stroma. Inasmuch as the principal medium of communication between these two tissues is the so-called *extracellular matrix* (ECM) to which they both contribute and with which they both attach and interact, the main focus of the Bissell laboratory has been the relationship between mammary epithelium and the ECM (Werb et al. 1996). Mammary cells (as well as other types of epithelium) attach to ECM by way of cell-surface attachment proteins, especially members of the "integrin" family of receptors. Such attachments also transmit signals to the interior of mammary cells which then may affect many processes, including transcriptional state, and ultimately the way in which these cells in turn modify (degradatively and/or synthetically) the ECM. Epithelial-based changes in the ECM will induce reactions in the cells of the stroma leading to a new volley of reciprocal cellular activities. What Bissell and others have observed to be a common feature of all breast cancers (and many other types of cancer as well) is an abnormal cell-to-ECM relationship (Boudreau & Bissell 1998). Both in vitro and in vivo studies by Bissell as well as others provide evidence to the effect that an abnormal ECM can lead to the initiation and/or promotion of mammary tumorigenesis and that the right ECM can stabilize a nascent carcinoma (Radisky et al. 2001). Bolstered by such findings, Park, Bissell, and Barcellos-Hoff (2000) have anted up the following questions with a strong anticipation of affirmative answers in the making.

1. Can changes in the microenvironment precede the progression of neoplastic disease?

2. What features of the microenvironment promote neoplastic disease? Are these tissue-specific?

3. Can the microenvironment be targeted therapeutically to prevent cancer?

4. Can manipulating the microenvironment reverse cancer?

If and when an appreciation for the role of the microdynamics of cellular ecology in tumor formation *and metastatic progression* becomes central to the oncological research paradigm, a major shift of perspective toward what I have earlier referred to as a "new epigenetics" will have been realized.

From Somatic Mutation to Genetic Predisposition—the Paradoxes of Progress

Molecular analysis is not needed to distinguish what percentage of human cancers are based on genetic inheritance—classical epidemiology provides that information. Either cancer is seen to follow familial patterns of inheritance and is thus putatively heritable, or it is seen to arise sporadically. Epidemiological studies have consistently found the vast majority of cancers to be sporadic in nature, prompting oncologists to see somatic as opposed to germ-line mutations as of more relevance to the understanding of cancer etiology.

Why then has so much recent attention turned toward questions of genetic susceptibility? Have the past appraisals of the role of inheritance in cancer morbidity been misguided? And what role has the succession of putative breakthroughs in understanding the role of somatic mutations had in leading up to the new turn toward germ-line genetics? This last section will address these questions and offer an answer that may be surprising.

The first part of this discussion will focus on a rather remarkable and revealing scientific review entitled "Lessons from Hereditary Colorectal Cancer" by Kinzler and Vogelstein (1996). The Kinzler and Vogelstein review is of much interest for a number of reasons. In the previous section I have highlighted the distinction between the center, or mainstream, and the margins or periphery in the history of twentieth-century cancer research, identifying the ongoing effort toward realizing a somatic-mutational, and thus subcellular, basis of causation as the mainstream. By 1990, as much attention shifted away from oncogenes to tumor suppresser genes, Bert Vogelstein was already well established as a leader in the field by virtue of his work on the p53 tumor suppresser gene. His papers have been ranked consistently during the last 10 years among those most frequently cited. Most importantly, Vogelstein's use of human colorectal cancer as the basis for describing a multistep, multimutational progression has become the principal exemplar by means of which the somatic mutation model of cancer has come to be understood. There can be little doubt that Vogelstein's name is perennially present during the annual deliberations of the Nobel Prize committee. The Kinzler and Vogelstein review appeared in the periodical *Cell*, a journal that has enjoyed the status of being the most prestigious among molecular biologists for more than 20 years. The very fact of having a review article published in *Cell* is itself an unmistakable indication of high status and recognition. The Kinzler and Vogelstein review is thus as mainstream as it gets.

Kinzler and Vogelstein set out to detail the role of two rare germ-line mutations in the overall etiology of colorectal carcinogenesis in relation to the schema of sequential mutations, which has become well known to every student of sporadic cancer. The article ends with an endorsement of genetic testing for colorectal cancer susceptibility along with an espousal of ethical, legal, and social considerations that need to go along with it. Kinzler and Vogelstein have thus spanned the distance from the disciplinary space in which the most heady proclamations of victory over cancer in the 1980s were heard, all the way to the prudently actuarial stance of the 1990s genetic counselor. Why such a strategic retreat? A close look at their analysis of the genetics of colorectal cancer, in the company of the insights of previous sections, will provide the answer. The key issue will be whether mutations (somatic and/or genomic) are driving the phenotype, as the somatic-mutation hypothesis must require, or whether in fact the phenotype, understood in terms of cellular behavior and its interactions in a microenvironment, is driving the conditions of mutability of its genome.

Colorectal cancer is the most prevalent cause of cancer mortality in the United States when smoking-related cancers are excluded. At least 50 percent of individuals in the West develop a colorectal tumor by the age of 70, and in about 10 percent of these progression to malignancy ensues. Epidemiological studies have identified about 15 percent of colorectal cancer incidence with patterns of dominant inheritance. In classical terms this would suggest that there are certain single genes which can cause colorectal cancer in a dominantly acting fashion. Yet many studies on colorectal cancer by Vogelstein and others have established that at least seven mutations are required. What has distinguished the colorectal model in particular is evidence of specificity in the order in which these mutations must appear for carcinogenesis to ensue. There thus appears to be a correlation between certain precancerous stages of dysplasia and neoplasia with certain mutations that open the door to the next stage. The sequence has been characterized as follows: The initiator mutation is the APC gene, in which both alleles must be knocked out. This results in the formation of many benign "aberrant crypt foci." Of these a small number may progress to the early adenoma stage, at which time the activation of a single k-ras gene is required for further progression to the intermediate adenoma stage. The late adenoma stage is attained by way of additional mutations on the 18q21 chromosome, such as that of the DCC gene, again requiring knockouts of both alleles. Finally, the transition to a frank carcinoma is reached by way of the loss of both p53 alleles. Further progression to invasiveness and beyond would presumably require additional lesions (Kinzler & Vogelstein 1996). The unavoidable problem that a model such as this raises is that of how so many mutations, each of which is presumably a very low probability event, can possibly occur during a time frame relevant to a human life span. Recent findings on the genetics of familial colorectal cancer provide additional resources for considering the matter.

The two forms of hereditary colorectal disease that have been best characterized are termed FAP and HNPCC (familial adenomatous polyposis and hereditary nonpolyposis colorectal cancer, respectively). The specifics of these will be examined in order.

FAP is the result of the inheritance of a mutation in an APC allele. While germ-line mutations of APC only account for less than 1 percent of colorectal cancers in the United States, FAP does provide a window onto what has been characterized as the first step in the colorectal carcinogenesis pathway. Patients with the single germ-line APC mutation typically develop hundreds to thousands of colorectal polyps by their second and third decades of life. Every colorectal cell of the FAP patient does not give rise to a polyp, but rather only about one in a million. This would be consistent with a requirement for a second mutation knocking out the wild-type allele. The effects of the APC mutation are highly tissue-dependent. APC is expressed ubiquitously throughout the body, and yet the mutations are not known to be associated with cancer in other organs. Nor do the same mutations necessarily result in the same clinical symptoms amongst FAP patients; nor does FAP necessarily lead to colorectal carcinoma, although it certainly increases its probability. In the overall incidence of colorectal carcinoma, the APC mutation has been thus far identified in about 80 percent of cases. While the ras and p53 mutations have been identified as necessary for further progression toward full-blown colorectal cancer, neither of these mutations shows any correlation with colorectal cancer in the absence of an APC mutation already being present. So it would appear that in the specific context of colorectal epithelium (and only there), the APC mutations, while not sufficient for causing cancer, are necessary in the large majority of cases

(Kinzler & Vogelstein 1996). What then is it that the APC mutations can be said to cause?

The sequence of the APC gene does not provide any clues by way of analogy as to the function of the APC protein because it does not show homology with other genes or proteins that have been characterized. Where inferences with respect to APC function are to be found is with respect to the binding of the APC protein to other proteins. The central third of APC contains two classes of binding sites for the protein β-catenin. The catenins are cytoplasmic proteins that bind to the family of cell-adhesion molecules known as *cadherins*. Evidence suggests that binding to β-catenin is necessary for the cadherin to function in binding adjacent cells together. Given that the binding of β-catenin to cadherin and its binding to APC are mutually exclusive (i.e., competetive) (Kemler 1993), APC may function to modulate the ability of the colorectal cells to bind to adjacent cells. In addition, the binding state of β-catenin will also affect the transmission of extracellular signals through the plasma membrane and into the interior of the cell (a process known as *signal transduction*).

What can best be surmised about APC is as follows. In the absence of either of the normal alleles, a protein is produced which results in the destabilization of cell-cell interactions in the colorectal epithelium, giving rise to an altered micromorphology known as *polyps*. The fact that these arise only in colorectal tissue, despite the ubiquitous expression of APC, suggests that it is only in the context of the specific biochemical-organizational state of differentiated colorectal tissue that this sensitivity to the status of the APC molecule can be found. The formation of polyps is a shift in the micro-organizational field of the colorectal tissue in response to a perturbation. In this case the perturbation is a change in cell-cell structure due to the alteration of an internal component which is involved in the architecture of cell-cell adhesion and signal transmission. Yet by analogy to the example of liver nodules discussed above, the colorectal polyps may also be a kind of adaptive response to any agent which presents a challenge to local tissue organization, whether this agent is internal and inherent or external in origin. And as in the case of the liver nodules, the colorectal polyps are not themselves cancerous but are just more prone to becoming so than normal tissue.

What constitutes then the first step in the direction of colorectal cancer is not the APC mutation as such but rather an alteration of the organizational field of the colorectal epithelium into the form of polyps. As already stated, it is only in the context of the formation of polyps that all the subsequent mutations associated with the onset of colorectal cancer are even relevant to it, as there is no correlation otherwise—which is why the inheritance of FAP is seen to have an apparently direct relationship to colorectal cancer. The next question is what might be the relationship of polyp formation to the subsequent stages of carcinogenesis in general, and the acquisition of the ras, DCC, and p53 mutations in particular? It turns out that this matter becomes best addressed in the context of the HNPCC findings, and with some surprising results.

HNPCC, as a syndrome, simply reflects the epidemiological observation that there are incidences of elevated levels of colorectal cancer which do not involve polyposis but do appear in familial patterns consistent with Mendelian inheritance. Only during the last several years did several lines of evidence come together to reveal that the inherited defects were in enzymes associated with the repair of errors in DNA copying. Errors in copying are most likely where there are regions of DNA with noncoding repetitive sequences. These regions are referred to as *microsatellites* and the failure to be able to repair errors of microsatellite sequence during DNA replication is observed as *microsatellite instability* (MI). HNPCC was thus shown to involve inherited lesions to mismatch repair genes (MMR), resulting in cells with microsatellite instability and a mutation rate two to three orders of magnitude higher than normal cells. HNPCC accounts for only 2 to 4 percent of colorectal cancers in Western countries. Another 13 percent of total colorectal cancers were also found to have microsatellite instability, and some of these reveal sporadic (as opposed to heritble) mutations of the same MME genes. The remaining 85 percent of colorectal cancers do not show microsatellite instability or heightened mutability, and yet they display gross chromosomal losses, with an average loss of 25 percent of randomly selected alleles. Colorectal cancers thus fall into two distinct categories. One group displays a heightened rate of mutability and small-scale genome-wide sequence errors (MI), but they do not show losses of whole alleles. The other group does not test positive for increased mutability and does not show MI but has lost at least 25 percent of its genomic alleles. At the cytological level, the large-scale loss of alleles is referred to as aneuploidy.

Aneuploidy is a common feature of all sorts of cancers at more advanced stages of progression. The mechanisms resulting in aneuploidy are not known, but it has generally been treated as a kind of side effect of various aspects of the carcinogenic process as opposed to a specific physiological state with specific causal significance. Kinzler and Vogelstein (1996) are moved to wonder:

If aneuploidy were simply the consequence of the neo-plastic factors described above [numerous extra cell divisions, abnormal microenvironment, altered physical structure of the cancer cell], it should be found in colorectal tumors with MMR deficiency as often as in other colorectal tumors.

This conundrum prompts Kinzler and Vogelstein to utter the following speculation which would appear to turn the fundamental presuppositions of the somatic mutation hypothesis on their head:

Teleologically, it would thus seem that a cancer needs to develop only one type of instability and the gross chromosomal changes provide little selective growth advantage to tumors with mismatch repair deficiency and vice versa. It would also seem that there are two ways for a tumor to develop the multiple genetic alterations required for malignancy: subtle alterations due to the mismatch repair deficiency occur in a minority of cases (those with MI), while gross chromosomal alterations occur in the majority.

Now whether it is really the intent of Kinzler and Vogelstein to suggest that the acquisition of gross chromosomal changes is not passively obtained but rather the result of active, adaptive processes is not even pertinent, given that there is no better explanation on the table (which they have set). Consider again the initiation of colorectal cancer with the formation of a polyp. The major problem was to account for how all the mutations which are required for full-blown cancer could possibly come about, assuming that mutations are passively-acquired, low-probability events. But another picture has now presented itself. Perhaps mutability of the cell is physiologically determined, a function of its status in an intercellular field. The polyp already constitutes a kind of disturbed field as did the liver nodule. Neither most nodules nor most polyps progress to malignancy, but under certain conditions some do. This progression entails a widespread genomic reorganization brought about by way of the cell's own resources, a reorganization (or loss of organization) which becomes inseparable from the cancer cell's ability to escape the constraining aspects of the tissue matrix.

Further evidence of the active role of a cell in modulating its own mutability in a disturbed field is provided by Kinzler and Vogelstein in examining the role of environment in colorectal carcinoma:

> There is little question about the importance of diet in limiting colorectal cancer incidence in the Western world. It has been a reasonable assumption that the dietary components responsible were mutagens. However, examination of mutational spectra in colorectal cancers has provided little evidence to favor specific mutagens as causative agents. The most characteristic mutations observed in p53 and APC genes, for example, are C-to-T transitions at CG dinucleotides (Harris and Hollstein, 1993). Such mutations are characteristic of endogenous processes leading to the hydrolytic deamination of methylated C residues in the absence of mutagen exposure. . . . Thus, it is possible that the dietary factors which lead to colorectal cancer are not mutagens, but rather irritants that lead to tissue regeneration. Dietary fibers may absorb these irritants, explaining part of their protective effect.

The logical prediction following from the somatic mutation hypothesis would be that mutagens in the gastrointestinal (GI) tract would be the leading environmental factor in promoting-initiating colorectal cancer. The evidence suggests otherwise, prompting Kinzler and Vogelstein to recover the concept of an "irritant," which harks back to the language of Virchow and company at the end of the nineteenth century when the adaptive capacity of a cell was still at the center of thinking. Kinzler and Vogelstein are unequivocal in the inference they draw from the mutational data: *Such mutations are characteristic of endogenous processes.*

In chapter 3, under the discussion of chromosome marking as an epigenetic inheritance system, the point was made that developmental and context-sensitive processes by which cells methylate and demethylate the C residue of CpG couplets may play a role in genomic reorganization as well. The example of colorectal carcinoma, in which the formation of an abnormal organizational field (the polyp) may lead to heightened mutability and eventually widespread loss of heterozygosity, may be just such a case. On close consideration it becomes apparent that the upshot of the colorectal multimutation model of carcinogenesis—very much the heir to the mainstream of the somatic mutation hypothesis—has come to dovetail with the story that has been told "from the margins" by the likes of Smithers, Farber, and Rubin. Genetic changes are tantamount to tumor progression, but the process is hardly unidirectional. Epigenetic systems of the cell (as discussed in chapter 3) are responsive to the organizational field of the cell and in turn influence the state, status, structure, and mutability of DNA. Enzyme-mediated processes are responsible for deleting, duplicating, amplifying, cutting, pasting, and relocating DNA all the time.

Cancer is about the protracted disturbance of all of those systems that stabilize a cell in its developmental compartment. There are evidently many ports of entry when it comes to initiating a disturbance and most likely none is ever sufficient in itself to dictate an entire carcinogenic trajectory. Chapter 3 argued that the informational content of the genome is co-original and reciprocally dependent upon that of the organizational structure of the cell (on all three levels of its epigenetic inheritance systems). Likewise, the genome and the organizational state of a cell are inseparably co-determinative of any aberrant departure from the developmentally stabilized status of a cell. Either hydrophobic environmental irritants that become lodged in cell membranes or mutation of cell-adhesion-molecule genes may perturb cell-cell interactions and cellular fields, which may lead to a stepwise uncoupling of cells from the tissue matrix, progressive destabilization, the reorganization of cellular genomes, and ultimately the death of an organism. The metastatic cell that invades a host organ and thus appears to be autonomous is the very late end product of a long, complex, interactive, and highly contingent process. Similarly, the transformed NIH 3T3 cell in Harry Rubin's laboratory, which outgrows other colony producers, had its ancestral origins in overcrowded growth conditions in tissue culture. It is hardly a cell whose fate was predetermined and dictated from within its genome. It may be that cancer has many potential ports of entry, albeit with none of them being sufficient to determine a carcinogenic trajectory.[8]

Now here comes the twist. Twentieth-century biology has been guided largely by the heuristics of some form of genetic preformationism. Cancer biology, going back to Boveri, has endeavored to explain cancer in terms of the genetics of somatic cells that have sustained mutations because the epidemiology of human cancer has never been consistent with the bulk of cancer incidence, being based on the inheritance of a genotype. Why then, at the end of the twentieth century, when the pre-formationist assumptions of even the *somatic* mutation hypothesis are being progressively undercut, has a *genomic* model of heritable suscep-tibility to cancer emerged and even moved onto center stage? Highly touted genetic tests for cancer, such as that developed and marketed by Myriad Genetics for the BRCA1 and BRCA2 genes for breast cancer, have contributed to a public perception that everything turns out to be genetic.

Has gene-based heritability proven to play a greater role in the etiol-ogy of cancer than previously suspected? Certainly not. Again, the example of colorectal cancer provides a model for what is at least equally the case for breast cancer. The identification of the heritable germ-line mutations associated with no more than 15 percent of colorectal cancers provides clinicians (and drug companies) a place to look, because just as for the proverbial drunk who's lost his keys and looks for them beneath the lamppost, it is where the light is. Cognitively speaking, the move from *somatic mutation* to *genomic susceptibility* represents a retreat toward a new burst of instrumental preformationism in the face of real advances in the understanding of the complex epigenetics of carcino-genesis. In the larger sociocultural context it reflects an unprecedented influence of the marketplace on the biomedical research agenda as well as on the correlative state of public understanding.

5

After the Gene

Based upon their fundamental roles in genome transmission and in determining patterns of gene expression, it can be proposed that repetitive DNA elements set the "system architecture" of each species. . . . From the system architecture perspective, what makes each species unique is not the nature of its proteins but rather a distinct "specific" organization of the repetitive DNA elements that must be recognized by nuclear replication, segregation, and transcription functions. In other words, resetting the genome system architecture through reorganization of the repetitive DNA content is a fundamental aspect of evolutionary change.

—James Shapiro, 1999

• There appear to be about 30,000–40,000 protein-coding genes in the human genome—only about twice as many as in worm or fly. However, the genes are more complex, with more alternative splicing generating a larger number of protein products.

• The full set of proteins (the "proteome") encoded by the human genome is more complex than those of invertebrates. This is due in part to the presence of vertebrate-specific protein domains and motifs (an estimated 7 percent of the total), but more to the fact that vertebrates appear to have arranged pre-existing components into a richer collection of domain architectures.

—International Human Genome Sequencing Consortium, 2001 (Lander et al. 2001)

Although we still lack the analytical tools, there is a growing appreciation that organisms constitute complex, self-organizing systems whose properties can be understood through the study of interactions within and between networks of mutually interacting components, be they DNA sequences, proteins, or cells. Organisms must also be appreciated as historic entites. . . . Whereas a high level of internal redundancy is appreciated as one of the most distinctive features of the complex genomes of higher eukaryotes, the theoretic and practical difficulties associated the with origin and maintenance of redundancy, in my view, have gone largely unrecognized and may be central to understanding contemporary

genome structure. . . . A rapidly growing body of data from genome characterization, cloning, and sequencing in a variety of organisms is making it increasingly evident that transposable elements have been instrumental in sculpting the
contemporary genomes of all organisms.
—Nina V. Fedoroff, 1999

The science of life gyrates to a centennial beat. In 1800 it was first chris-
tened with the name "biology."[1] In 1900 it "rediscovered Mendel"
(Carlson 1966) and took its "phylogenetic turn" (see chapter 1). In 2001,
the preliminary findings of the Human Genome Project were reported
(Lander 2001, Venter 2001), constituting certainly the culmination
and, I would suggest, conclusion of the "century of the gene."[2] Comparative analyses of the human genome with that of the previously
sequenced fly (*Drosophilia*) and worm (*C. elegans*) genomes, bring into
striking relief realizations that have been bursting forth for some time
and provide the grist for some concluding comments on behalf of the
next gyration.

I suggested earlier that the gene-centered perspective was built of a
conflation of two individually warranted but mutually incompatible
conceptions of the gene (Gene-P and Gene-D) and that these were held
together by the rhetorical glue of the gene-as-text metaphor. And central
to the gene-as-text metaphor is the understanding that the biological
function of DNA is that of "coding." Much of the debate between contemporary preformationists (gene-centrists) and advocates of a new epigenesis can be construed as a debate about the *scope* of coding.

My Gene-D is not denied a special template (coding if we must)
function, but the scope of this function is limited to within an always
phenotypically indeterminate molecular level. Advocates of genetic
preformationism, by contrast, argue (by conflationary sleight of hand, I
argue) for a large-scale scope of coding, described as a genetic program,
book of life, and so forth, that determines the phenotype.[3] In either case
this debate looks at DNA *qua coding*, which is to say DNA *qua gene*.
At this latest biological *fin de siècle*, DNA has come to burst the bounds
of the gene itself. On the threshold of the "postgenomic" era it is has
become possible to glimpse ahead to the nature of molecularized biology
after the gene.

Modularity, Complexity, and Evolution

Comparisons of the human and invertebrate (fly and worm) genomes have come to reinforce certain growing realizations that are reflected in the epigraphs above. These realizations have to do with biological modularity and its relationship to organizational complexity; with the dynamism of the genome; and with the significance of repetitive non-coding "parasitic" DNA to both of the above. Once upon a time it was believed that something called "genes" were integral units, that each specified a piece of a phenotype, that the phenotype as a whole was the result of the sum of these units, and that evolutionary change was the result of new genes created by random mutation and differential survival. Once upon a time it was believed that the chromosomal location of genes was irrelevant, that DNA was the citadel of stability, that DNA which didn't code for proteins was biological "junk," and that coding DNA included, as it were, its own instructions for use. Once upon a time it would have stood to reason that the complexity of an organism would be proportional to the number of its unique genetic units. Beginning with the discovery that eukaryotic genes are assemblages of ancient modules (Gilbert 1978) and with recognition of the actual dynamism of DNA a very different picture has progressively emerged.

The percentage of the human genome which is responsible for protein coding is extremely small (less than 1.5 percent) (Baltimore 2001). It is organized into modules referred to as *exons*. The exons themselves tend to be highly conserved throughout phylogeny going back to the one-celled stage. Exons generally correspond to a domain of a protein, a domain being a piece of protein that has some structural or functional integrity. The ability of proteins to bring specificity to chemical reactions, whether in catalyzing a reaction, forming durable structural elements (filaments, muscle, etc.), transmitting signals, or binding an "antigen" is localized to the specificity of its domains. Human genes (Gene-D) on the average consist of 7 exons but may vary from as few as one to as many as 178 (Lander et al. 2001) and thus code for an average of 7 (but as many as 178) possible domains. Typically the exon modules are dispersed within the transcriptional unit (i.e., the Gene-D), like islands within a sea of much more extensive intervening sequences (introns).

Neither humans nor higher organisms in general are distinguished by their repertoire of exons. The sum total of all bacteria contain as many, and almost certainly far more, kinds of exons than the sum total of all multicellular (and one-celled eukaryotic) organisms. Accordingly, bacteria display a far greater range of metabolic versatility than the sum total of all higher organisms. There is scant evidence for evolution being built upon the expansion of the number of basic coding modules.[4] Genes (Gene-D) are composed of assemblages of old modules, and the increase in the number of genes one sees in going from bacteria to one-celled yeast to invertebrate to humans is based on the greater number of ways in which modules have been sorted into different combinations.

Humans have about twice the number of genes (Gene-D) as fly and worm but less than 7 percent of this difference is accounted for on the basis of apparently novel domains. The remainder is due to the further regrouping of exon modules in the genome. But the difference in Gene-D number between humans and these invertebrates does not account for the difference in the complexity of such organisms.

The evolution of increasingly complex organisms, it turns out, is based upon the evolution of increasingly modular architectures. The critical decisions made at the nodal points of organismic development and organismic life are not made by a prewritten script, program, or master plan but rather are made on the spot by an ad hoc committee. And these committees consist of ensembles of modular parts, the composition of which are contingent upon circumstance. And the more complex the organism, the greater the number of different potentially modular constituents and the more sensitive is the outcome to the nuances of circumstance.

Gene-D is built out of modules (exons). The modular architecture of a Gene-D allows for expanding the array of Genes-D through shuffling the modules into new configurations. But more importantly the modular architecture of a Gene-D provides for great flexibility and variablity in how the Gene-D, as a resource for making a protein, is put to use. Consider the example of the Gene-D called NCAM. NCAM contains 19 modular exon units (figure 5.1) but there are no NCAM proteins that are composed of the protein domains coded for by all 19 exons. Any NCAM protein is the result of some subset of these potential domains,

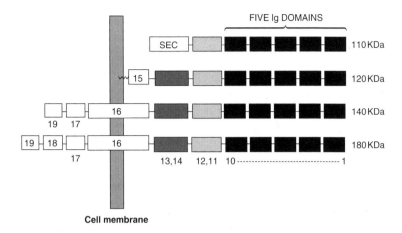

Figure 5.1
A schematic diagram of the four main classes of NCAM protein; see text for details.

and many different subsets are possible. The process by which a particular configuration of modules is assembled is called "splicing." It is not the DNA but rather a messenger RNA transcript that includes the complementary RNA version of all of the exons that become subject to splicing. The ensemble of possible NCAM forms is classified into four main groups depending upon size (110 KDa, 120 KDa, 140 KDa, 180 KDa) and plasma membrane attachment (see figure 5.1). Whereas the 140 KDa and 180 KDa classes traverse the plasma membrane, the 120 KDa class is linked to the membrane only superficially through an auxiliary connector.

No N-CAM does both. And the basis for this difference is the either/or inclusion of one of two different exon modules into the final RNA transcript. The 140 KDa and 180 KDa must be derived from a transcript which includes exon no. 16 in order to be able traverse the plasma membrane, but must lack exon no. 15, whereas the 120 KDa NCAM class is derived from a transcript which must include exon no. 15 in order to be able to associate by auxiliary connector, but must lack exons nos. 16 to 19.

The human genome has twice the number of Genes-D as that of fly or worm, but the human *proteome* (the full set of all expressed proteins) is thought to be at least 5 times as complex as that of invertebrates. This is because of an enhanced variability produced through transcript splicing. For just the case of a single Gene-D such as NCAM—and even prior to questions of post-translational compartmentalization (see chapter 3)—there are two regulatory nodal points that determine its fate: transcription initiation and splicing. Both of these are adjudicated by the complex proceedings of "ad hoc committees."

Whether a given Gene-D becomes transcribed into RNA to begin with is determined by two categories of proteins: the transcriptional factors, including the polymerase enzyme, which have been highly conserved from yeast to humans, and the transcriptional effectors whose ranks have expanded over evolutionary distance. Back in Chapter 3, the example of an artificially simple, hormone-sensitive proliferin transcription system was used to show how complex the effects of even just three transcriptional effectors can be. The regulation (*yes* or *no* and *how much*) of Gene-D transcription is determined by an ad hoc committee—which is to say it is a function of the complex relations of all of those transcriptional effector constituents present in the nucleus at that time.

Human Genome Project findings suggest that the human genome possesses 2000 transcriptional effector genes (over 5 percent of the entire gene number), a major increase over the number found in the other sequenced species (Tupler et al. 2001). Inasmuch as the role of each transcriptional effector depends on the identity of as many as 2000 other effectors, the complexity of the transcriptional initiation event and its potential sensitivity to ancillary events that influence the composition of the "ad hoc committee" is enormous.

What is true of transcriptional initiation is also true of transcript splicing. The ad hoc committee that determines how an RNA transcript will be spliced (and thus just what biological significance the resulting protein might have) is called a *spliceosome*. Its effects are a complex function of composition and assembly, and again the human genome reveals a significant expansion in the number of potential committee members (Tupler et al. 2001).

The evolution of complex, internally differentiated, and yet globally coordinated life forms, including *Homo sapiens*, was achieved not by the elaboration of a master code or script but by the fragmentation of the functional resources of the cell into many modular units whose linkages to one another have become contingent (Gerhart & Kirshner 1997).

The more contingently uncoupled the molecular and multimolecular constituents of a cell become, the greater becomes the subset of potential specializations that can be achieved. The decision as to when to couple or not is made, as we've seen, by ad hoc committees. And the roster of potential committee members has decisively expanded in concert with the increasing complexity of an organism.

Why might this be? Might it be the case that what has underwritten evolutionary complexification is not the expansion of the number of enyzmatic craftsmen but rather the number of molecular "regulatory lawyers and politicians" who adjudicate at the coupling-uncoupling nodal points. The character of a cell, its differentiated cellular identity, is generally correlated with the particular set of proteins of which it is composed. Transcriptional initiation and splicing are the first two nodal decision points that determine the composition and, to a large extent, identity of a cell. The committees that adjudicate these processes can be thought of as constituent assemblies. Each constitiuent in turn reflects some set of enabling conditions. The presence of a constitutent reflects not only the past history of the interior of the cell but also the recent history of interactions with other cells and the extracellular environment. The more complex the constituent assembly, the more facts about the past and present history of the cell and its surround are being brought to the decision-making table. The evolution of complexity is the evolution of increasingly sophisticated levels of horizontal and vertical coordination.

The evolutionary expansion of the ranks of representative intermediaries appears to be central to the means of achieving this. What each ad hoc committee does in regulating transcription and splicing is to achieve a kind of consensus about the "state of things." A larger ensemble of potential constituents means a wider sampling of news from the hinterlands—a richer cellular *Umwelt*. The state of each cell then becomes better coordinated with that of the tissue, the organ, the organism as a whole, and so on. In addition, a larger more differentiated committee allows for buffering. The consensus of a complex committee can be such that the absence (or mutation) of any one constituent need not be decisive. The ad hoc constituent committee simultaneously expands the reach of causal influence and yet dampens the effects of any one particular influence. Two phenomena discussed in the previous chapter can be better understood in the light of this analysis. The notion of an intercellular field and its influence on constituent cells becomes more palpable with the realization that many intracellular processes, mediated by "complex constituent committees," serve as causal funnels for bringing a great variety of ambient influences to bear on intracellular events. Cell-cell contacts, cell-matrix contacts, ionic characteristics, receptor-mediated events, even steric constraints, can all influence the composition of the constituent assemblies with consequences that can rebound back to those regions of influence. Second, and not unrelated, is the context sensitivity of somatic mutations and the loudly heralded discovery of redundancy. For 100 years somatic mutation theorists have wanted to affix the causal basis of malignancy to a purely internal condition of the cell. But if cells are not dictated by an internal script but rather by ever so many ad hoc committees whose constituents reflect the dynamic state of the cell and its larger milieu, then this inside-outside dichotomy is rendered bogus. Likewise, the sudden absence or aberrance of a transcription-initiation committee member, such as p53 (chapter 4), may have a grave impact on cellular behavior in the specific context of a certain tissue, intercellular field, and constituent assembly, while the same molecule (p53) when missing from birth (as engineered in transgenic mice) may have negligible impact because its absence was accommodated through "constituent buffering" from the get-go, resulting in a systematically modified developmental history.

If the sum total of coding sequences in the genome be a script, then it is a script that has become wizened and perhaps banal. It isn't the script that continues to make life interesting but rather the ongoing and widespread *conversations about it.*

What DNA Can Do

If it were the case that genes coded for all the information needed to build anything from a yeast to a fly to a human being, then the idea that the vast majority of the genome—all that isn't involved in coding—is merely junk might be tenable. But if indeed it is primarily the *regulation* of what boils down to the same old stuff that evolves, and if organismic complexity is built by pulling apart the pieces and expanding the range of choices to be made at the decision-making nodal points, and if these nodal points are fulcrums for wide ranges of influence, then the proponents of conflationary gene-speak have missed the forest for . . . well, one species of tree. Over half of the human genome consists of repetitive sequences and most, if not all, of these consist of parasitic transposable elements (Lander et al. 2001).

Transposable elements come in four principal varieties, three of which transpose through RNA intermediates and one of which transposes directly as DNA. Transposable elements are referred to as *parasitic* because they come well equipped with their own promoters, reverse transcriptase and other enzyme templates, for advancing their own replication within the genome. But as such they also represent a source of dynamism, of mobility, and of architectural innovation and reconstruction. Transposable elements, it now appears, have been the motor-force of genomic innovation from the one-celled stage onwards. It is these engines of activity that have created new genes, but more importantly have created regulatory binding motifs that many of the regulatory committees become targeted to, and even more importantly it appears to be transposable units that have served to modularize genomes (Fedoroff 1999; Lander 2001). Far from being biologically irrelevant, the spacing, positioning, separations, and proximities of different elements in a complex system of distributed regulation appear to be of the essence (Shapiro 1999), and it is "transposable elements (that) have been instrumental in sculpting the contemporary genomes of all organisms" (Fedoroff 1999). But how could this be? How can ostensibly parasitic DNA be essential to the evolution of higher, increasingly complex, multicellular life-forms?

Symbiotic Symphonies

The physicist Freeman Dyson (1999) has provided us with what is arguably our most interesting and plausible model for the origins of life and in so doing simultaneously solved our puzzle in advance. Dyson envisaged the origin of living cells taking place not in one but in two steps. In the first step a boundary-maintaining, autocatalytic, metabolizing system consisting largely of proteins becomes established. It is only after this, in the context of such ongoing metabolizing enterprises that RNA evolves as an intracellular, self-replicating parasite which eventually becomes symbiotically integrated into the life cycle of its host. The host provides the precursors for RNA replication and the RNA (and eventually DNA) comes to form through the evolution of translation mechanisms (and thus codons), an efficient repository of protein template information.[5] Dyson takes his clue from "the strange fact that the two molecules, ATP and AMP, which have almost identical chemical structures, have totally different but equally essential functions in modern cells. ATP is the universal energy carrier. AMP is one of the nucleotides that make up RNA and function as bits of information in the genetic apparatus" (Dyson 1999). Indeed ATP is a precursor in the synthesis of RNA. How could one molecule come to be so central to two entirely different roles? By Dyson's schema, ATP had become established in its metabolic energy-carrying role. Various cells built up large quantities of it (as many cells still do). In some cases it spontaneously polymerized into proto-RNA becoming, as it were, the first virus. Initially cells would have become sick and died of this parasitic disease. But then, following the pattern of endosymbiosis described by Lynn Margulis[6] for the evolution of the eukaryotic cell, some infected cell would have developed the means to survive the infection and turn the relationship into one of symbiotic mutualism.

Dyson's model has a number of benefits. Amino acids, the precursors of proteins, are known to be capable of coming about spontaneously in a prebiotic environment. Indeed, they have even been found on meteorites. Nucleic acids, like AMP, are much less likely to be spontaneously produced and much less stable afterwards if they are. Nucleotides would have a much better chance to accumulate and polymerize inside of a metabolizing host. But even more interesting (and unremarked upon by Dyson) is the heuristic perspective that Dyson's model suggests. That which began as a parasite because of its efficient self-replication capacity, brings along the perennial threat of new bouts of parasitism and new rounds of symbiotic accommodation. Nucleic acids (RNA and DNA) began as viruses and have never stopped giving rise to viruses. Host cells have long since developed defense mechanisms against viruses marked as "other," but what about new rounds of symbiosis?

Almost half of human DNA (ten times more than that which is "coding") is parasitic (Baltimore 2001). But it doesn't constitute a sickness. Why not? The uncontrolled expansion of transposable elements in the genome is prevented through methylation ("epigenetic" chromatin marking as discussed in chapter 3). And it is possible that the methylation system evolved for this purpose (Matzke et al. 2000; Symer and Bender 2001). Eukaryotic cells have developed the capacity to recognize repeat DNA and to prevent or regulate its transpositional spread through methylation. By the terms of the latest symbiotic *modus vivendi* (literally) parasitic DNA had become incorporated into the living enterprise, not for template-code stabilization but as a source of controlled *destabilizaton* for modularizing, complexity promoting, architectural restructuring of the genome.

Evolution involves an ongoing symbiotic interplay between metabolic hosts and perennially short-circuiting segments of nucleic acid. The creativity and volatility of the enterprise is realized in the interlocking dynamics of the systems of DNA repair, DNA recombination, DNA transposition, and DNA methylation (Fedoroff 1999). Turn on the faucet and reconfiguring forces begin to flow. Turn it on faster and, as Fedoroff suggests, either evolution or cancer may ensue. Bacteria have been shown to turn on their "faucet," referred to as an SOS system (Radman 1973; Radman 1999) in response to conditions of stress. Metastatic cancer cells may well be reconstituting their own SOS system in (dynamic) response to becoming uncoupled from the stabilizing influences of a tissue matrix.

From bacteria to humans it appears to be a universal ability of living cells and living organisms to be able to turn on the faucet and allow some destabilizing forces to flow. One may want to question whether and how this may pertain to evolutionary dynamics that must cut across the Weismannian barrier between the soma and the germ line (for those organisms, like us, that sequester their germ line at any early stage of development). Evidence for the plausibility of this model can be gotten from the unfortunate example of Huntington's disease. What turns a certain Gene-D into a Huntington's Gene-P is the inclusion of multiple copies of the CAG trinucleotide (resulting in the inclusion of polyglutamine tracts in the translated protein known as "huntingtin"). The severity of the disease appears to be correlated with the number of such repeats,[7] with the greater the number predicting an earlier age of onset. The finding that age of disease onset has had some tendency to decrease in subsequent generations has been referred to as "anticipation." What is revealing is that anticipation results from a cross-generational expansion of the number of repeats, especially when the gene is passed from father to child. Male and female gametogenesis produce different patterns of chromatin methylation (imprinting) in the genomes of the resulting gametes. Anticipation in the transmission of Huntington's disease thus has the earmarks of a process that involves the interlocking network of DNA modulating enzyme systems proposed by Fedoroff: DNA repair, recombination, transposition, and methylation. Could it be that Huntington's and other CAG repeat-related diseases are tragic byproducts of those systems of genomic architectural reconstruction that occasionally give rise to evolutionary novelties? In any case the example of Huntington's disease provides a strong indication that even in human gametogenesis there is no ironclad barrier against forces of genomic reconstruction.

Rebuking the Dawkinsonian Replicator

Talk about parasitic DNA and its continuing role in the evolution of lifeforms may give some the impression of lending credence to trendy talk about the selfish gene also known as "the replicator." It shouldn't. For Richard Dawkins and his epigones it is the parasite that invents the host (see table 1.1). Dawkins's selfish replicator constitutes the quintessence of conflationary confusion. His viewpoint does not build on the advancing elucidation of molecular biology but rather depends on an enforced ignorance of it. His is a biology built of ontotheology. His point of departure conditions all that follows. Dawkins and his followers take their conflationary replicator (the so-called selfish gene) as an ontological bedrock. The "fact" of the primacy of the selfish replicator that stands in a creator-created relationship to all else is simply asserted as the grounds of all possible biological intelligibility. The Dawkinsonians then weave out the logic that follows from their ontology, and they defend it with religious zeal.

But why buy into such an ontology to begin with? The idea that a naturalistic account of evolution within a roughly Darwinian farmework requires a fundamental replicator is patently false. The Darwinian tradition stakes its claim on the idea of descent with modification. Nowhere does the need for either Xerox machines or Turing machines follow from this. Now the idea of doing biology by way of ontologic has a strong appeal to bioideologues and philosophers who are good at logic and weak on biological details. But each time intellectually honest thinkers attempt to make good on a consistent and coherent notion of an evolutionary replicator they come to realize that boundaries become blurry, that whole organisms (Hull 1980) and even such things as nests and birdsongs (Sterelny et al. 1996) must also be classified as replicators, and/or that in any event replicators are not necessary for Darwinian evolution (Godfrey-Smith 2001).

Parasitic DNA is just that. It is an ability of some chemicals to proliferate by chain reaction given the right conditions. The parasite did not invent the host, and it has only been the conflation of two nonoverlapping senses of the gene that has made it even tempting to think otherwise. When nucleic acids enter symbiotically into the enterprise of the living system, they do so as one kind of resource among many. It is the whole enterprise that will sink or swim, fail or succeed to leave descendents that bear some resemblance, and inherit components and contextualize them in a structurally and dynamically similar fashion.

Evolutionary Developmental Biology (Evo-Devo) and the Developmental Basis of Evolvability (Devo-Evo)

Aristotle approached reproduction as a continuation and outgrowth of development ("generation") and set the tone for centuries to come. During the "century of the gene" however, the processes of biological change that take place during one generation and those across generations became radically dichotomized.[8] Transgenerational change (evolution) was deemed to be the result of random point mutations that take place, as it were, behind the backs of organisms and so are not teleo-logical in any sense. Development, by contrast has been deemed to be the result of a preset centralized program and so teleological (or "teleo-nomic"[9]). Evolution, by virtue of the invisible hand of natural selection, has been deemed to be adaptive. Evolution is thus seen as adaptive but not teleological, development teleological but not adaptive.

The asymmetry of development and evolution during the twentieth century was not a historical accident. The processes of life have been dif-ficult to explain in strict accord with the presuppositions of natural mechanism (of course this is subject to variation with changes in the per-ceptions of what counts as natural mechanism). Darwin provided the schema for a natural mechanism that can explain how the forms of life can change over generational time. Increasingly, neo-Darwinists have attempted to bring all of the ultimate explanatory challenges of life under the ambit of the Darwinian schema. Taken to the extreme, this has led to a biological ontology in the form of the theology it sought to replace. The radical dichotomization of development and evolution followed from the apotheosis of the (conflationary) gene that became the funda-mental ground substance of all life and the basis of life's intelligibility (and of life's ability to become intelligible to itself).

At the beginning of our new century this dichotomization can no longer be sustained. The evidence against this is unmistakable and unavoidable. Development and evolution are not only two sides of the same coin; they are virtually mirror images of one another. The mecha-nisms of evolution have become essential to understanding development and the mechanisms of development have become essential to under-standing evolution. At the dawn of the new century, no ontological rocks may properly be left unturned.

The symmetry, interpenetration, and interdependence of development and evolution can be seen on many levels. New genes (Genes-D) are evolved through recombining exon modules at the DNA level. Genetic diversity is obtained in development through the recombination (splic-ing) of exon modules at the RNA level. The de novo recombination events that allow the immune system to produce receptors for antigens never before seen mirrors in miniature the de novo recombination events that have resulted in new genomic architectures that may have been instrumental in the evolution of new species. The developmental capacities of metazoan life forms are at once the basis of their evolvability. Faucets of variability are deployed ontogenetically and by extension phylogenetically. Variability is mediated by the tribunals of N-many dispersed and distributed decision-making nodal points and by on-site, on-the-spot, modular committees. The stability and intelligibility sought for in idealized genes must be rediscovered in the complex dynamics of process. The evolving metaozoan capacity to mediate variability (from all directions) is the basis of increasingly adaptive ontogenetic plasticity. Modularization underlies the formation of contingent linkages between proteins, and between multi-protein complexes. It promotes increasing compartmentalization, increasing redundancy, and the growing capacity for exploratory behavior. These properties confer "robustness and flexibility on processes during embryonic development and in adult physiology" and at once "confer[s] evolvability on the organism by reducing constraints on change and allowing the accumulation of nonlethal variation" (Kirschner & Gerhart 1998). Ontogenetic adaptability and phylogenetic evolvability in the metazoa are reciprocal capacities—two sides of the same street. Kirschner and Gerhart (1998) focus on how robust, modularized, phenotypically flexible developmental systems allow for even random point mutations to have evolutionary clout through influence that can be magnified through mediation by the systems of contingent modular interaction. Even random point mutation becomes grist for the ontogenetic mill. "The consequences of mutation for phenotypic change are conditioned by the properties of cellular, developmental, and physiological processes of the organism, namely, by many aspects of the phenotype itself" (Kirschner & Gerhart 1998). What is true for point mutations (the likelihood of which are also largely regulated by the interlocking enzyme systems of DNA repair, recombination, translocation, and methylation) can only be at least as true for large-scale genomic reconstructions and indeed for extragenomic variation of any sort, the evolutionary consequences of which are equally subject to phenotypic interpretation and stable incorporation into reproducible life-cycles. Contrary to the last-ditch efforts of diehard gene-centrists, the developmental analysis of evolvability "Devo-Evo" cannot be segregated from the evolutionary analysis of development "Evo-Devo" (see Hall 2000).

The decay and demise of the gene as the bedrock of biological explanation and intelligibility will surely bring with it new explanatory challenges. That even individual cells can coordinate a multitude of highly contingent and quasi-independent decision-making processes merits some measure of explanatory humility. An example of a growing research program that may shed light some light at this level is that which follows from the realization that the nucleus also contains its own filamentous matrix. The scaffolding of the nuclear matrix may, along with histone modification (Forsberg & Bresnick 2001, Rice & Allis 2001, Jennwein & Allis 2001), provide the means of coordinating the decisions of ad hoc transcription and splicing committees (Berezney et al. 1995). But the nuclear matrix can be no new fundamental ground substance, no new ontological bedrock. A nuclear matrix may provide the on-the-spot, at-the-time solution to certain problems of decision making coordination, yet like its reflection in the cytoplasm, a nuclear matrix is itself the modularly variable product of a contingent dynamic history.

It may well prove to be the case that as the newly ontogenized understanding of evolution becomes more truly secular, our understanding of life (and perhaps of matter in general) will yet become more sacred. After the (conflated) gene, it is the living organism, an active agent of its own adaptive ontogeny and evolvability, that is once again poised to move back into the ontological driver's seat.

Notes

Chapter 1

1. "Shibboleth" is defined as a catchword or formula adopted by a party or sect, by which their adherents or followers may be discerned and those not their followers may be excluded.

2. The quote, from Brandon (1990), is not meant to suggest that this author is particularlary culpable; rather, it is meant as a general illustration of what had been taken as a point of departure for the earliest generations of Anglo-American philosophers working in the philosophy of biology.

3. See especially Aristotle's *Parts of Animals* and also his *Generation of Animals*.

4. For historical discussions of the continuity of Darwin's work and ideas with the ontogenetic tradition see M. J. S. Hodge (1985) and Robert J. Richards (1992).

5. See Maturana, H. and Varela, F. (1973).

6. Blumenbach's comment can be understood in more than one way. Richards (2000) argues that Blumenbach never understood the Bildungstrieb as a "heuristic idea" in a Kantian sense but rather was understanding it in a Newtonian fashion as an empirical force, the cause of which would always remain hidden and ultimately unknown. Blumenbach quoted Ovid in this regard—"causa latet, vis est notissima" [the cause is hidden, the force is well recognized]. Blumenbach acknowledges, following Kant's commendation of him, that the Bildungstrieb brings together the mechanistic with the teleological, but Richards suggests that Blumenbach neither adhered to Kant's view nor even necessarily quite understood it. Lenoir interprets Blumenbach as an "emergent vitalist" who see the vital force, the Bildungstreib as emerging from the organization of matter (which in Kantian fashion we must take as a given), but Richards argues that for Blumenbach it is the Bildungstrieb which is meant to explain the origins of living organization.

7. *Keime* was routinely used as the German translation for the French "germes." Expressing the preformationist ideas of both Bonnet and von Haller, i.e., as

preformed parts, it should not be confused with the "emboîtment" model of preformed whole miniatures. *Anlagen*, which derives from the German word *legen* meaning "to lay out," is translated as "organizational layout" or "disposition." Kant is the first to bring the words *Keimen* and *Anlagen* together in this technical usage, first in his 1775–1777 discussions of race and then in a passage of A66 of the First Critique of 1781 (Sloan, 2001). In these texts the meaning of Kant's use of *Anlagen* is that of a native structuring capacity or aptitude which brings an epigeneticist sense to the more preformationist connotation of *Keime* (Sloan, 2001).

8. This teleomechanistic model of evolution in which ontogenetic adaptation is playing a driving role is reincarnated in the twentieth century in the name of Baldwinism and Waddington's "genetic assimilation" (see Gottlieb 1992). In chapter 3, I refer to the possibility that the dynamics of chromosome marking may provide a mechanistic basis for ontogenetically driven evolutionary adaptation. Lamarck of course provided another nineteenth-century model, but the label "Lamarkianism" having become a term of derision amongst neo-Darwinist has served mostly to polarize and to obfuscate real biological questions.

9. Quoted in (and translated from the fifth Scholium of von Baer's *Entwickelungsgeschichte* by) Frederick Churchill (1991).

10. The separation of the soma from the germ line in development is only found in "higher" organisms and so is itself a product of evolution, not a basic characteristic of evolution. See Buss (1987) for an excellent discussion of this.

11. The word evolution is derived from the Latin *evolvere* which means to "unfold or disclose." Its use in biology begins with that of seventeenth-century preformationists for whom development consists of the unfolding or scrolling out of parts always already formed. The first such use was by Albrecht von Haller in 1744. Through the eighteenth-century "evolutionism is synonymous with preformationism." See Richards (1992) for an historical reconstruction of the changing usage of "evolution" with an appreciation for the continuities of its meaning.

12. De Vries didn't speak of "macro" mutations, but by "mutation" he meant a change of structure sufficiently monumental such as can result in a new species. In the terms of current usage, De Vries's mutation would have to rank as a "macromutation."

13. Penetrance refers to the statistical frequency with which a trait is expressed at all given the presence of the genotype. Expressivity refers the *degree* to which a trait is expressed, given the genotype. These terms were introduced by O. Vogt in 1926 (see Sarkar 1999).

14. The Whitehead Institute for Biomedical Research acknowledged in a press release (5/14/2001) that only 1 to 5 percent of cloned animals survive to adulthood (see http://www.sciencedaily.com/releases/2001/05/010511073756.htm). Cloned animals and the placenta that nourish them are typically "dramatically larger" than normal couterparts, are frequently riddled with birth defects, and die hours after birth. The possible longer terms defects of cloning techniques on

longevity and long-term health cannot be discerned anytime soon as the first successfully cloned animals (such as "Dolly") are as yet a long ways from natural old age. (As this book is being prepared for press, national network news has reported that Dolly—the first cloned sheep—presently 5 years old, has been diagnosed with arthritis. There is no comment yet on the cause of Dolly's premature condition.) Where cloning technique relies on the juvenescent state of the donor cytoplasm, i.e., an egg cell cytoplasm, "cloners" have been less than forthcoming—to say the least about the potential difficulties of "rejuvenating" the donor nucleus.

15. To result in a phenotypic difference requires viability, amongst other things, therefore there are a great many possible deviations from sequence norms which do not show up as Genes-P because they are not compatible with viability at all, as well as a great many other possible deviations from sequence norms that do not show up as Genes-P because the organism develops and functions without noticeable difference. See discussion of transgenic animals in chapter 4 for further elucidation of this point.

Chapter 2

1. The current number of observed cystic fibrosis mutations can be found, along with a great deal of additional information, at the Cystic Fibrosis Mutation Data Base (http://genet.sickkids.on.ca/cftr/).

2. Schrödinger did not introduce the term "translation," and I agree with Lily Kay (2000) that the word itself follows from a linguistic-informational theoretic trend independent of Schrödinger; yet, I think it is still compelling that Schrödinger laid a conceptual groundwork which requires some new means of bridging a kind of chasm and that turning it into a semantic chasm becomes a possible way to go.

Chapter 3

1. A heterodimer is a molecular unit composed of two distinct components as opposed to a homodimer, which is composed of two of the same molecules. The components of a heterodimer are typically referred to as alpha and beta subunits.

2. For reference to the theoretical underpinnings of "dissipative systems" see Prigogine (1980) in the reference section. Amongst those working within a Darwinist framework and seeking to reconcile Darwinian evolution with the dynamics of nonequilibrium chemical systems Depew and Weber have been in the forefront (see Depew & Weber 2001, Depew & Weber 1998, Weber & Depew 1996, and Depew & Weber 1995).

3. The estimate of the number of kinsases in the human genome reported by Celera (Venter et al. 2001) is about half of that reported by Hunter (1995).

4. Human females have two X chromosomes in the nucleus of every cell, but in each cell one of the two is condensed into permanently inactivated "Barr bodies." The X chromosome (maternal or paternal), which is inactivated, is variable from cell to cell.

5. This quote is taken from the "executive summary" of Oyama, Griffiths, and Gray (2001).

Chapter 4

1. Evidence for this and other industrial sources of carcinogenesis was not well documented until a century later (Triolo 1965).

2. Aneuploid cells have an irregular number of chromosomes; they are neither consistently diploid nor haploid and are generally prone to be unstable.

3. Harris's point here is that in order to distinguish between a dominant and a recessive allele there must be two alleles present at the relevant locus. However, in an aneuploid cell one cannot do this because one cannot presume the presence of the second chromosome. This problem is further intensified by the propensity for unknown chromosomal-genetic alterations to take place during the transfection process.

4. Any region of DNA which is deemed to be a gene resides on one of two strands of the double helix. Opposite the gene on its complementary strand will be a sequence of DNA, which consists of all the complementary base pairs to that of the gene. This is what would be meant by the gene in the anti-sense configuration.

5. The codons for such an allele would of course always be present as they are just the complement of the codons of the fibronectin allele. What would constitute the formation of an anti-sense "allele" would have to be the acquisition of upstream promoter sequences that bind the polymerase complex and allow the anti-sense codons to show up as a transcriptional unit.

6. Soluble proteins released by white blood cells.

7. Mantle's father and grandfather had died in their early forties of Hodgkin's lymphoma.

8. A quite sympatico analysis can be found in *The Society of Cells* (Sonnenschein & Soto 1999) which can also be used as a kind of primer on carcinogenesis.

Chapter 5

1. The first use of the word *biology* was by Karl Friedrich Burdach in 1800 to refer to the study of man from a zoological perspective. In 1802 Gottfried Reinhold Treviranus and Jean-Baptiste Lamarck both used it with its current meaning implied (Richards 2000).

2. I take this apt description from the title of Evelyn Fox Keller's book (Keller 2000).

3. Perhaps the most rigorous and intellectually honest attempt to salvage a minimal preformationism based upon an analysis of the scope of coding is that of Godfrey-Smith (2000).

4. Consider for example that what enables the cow to digest grass and the termite to digest wood is not an evolved autochthonous capacity but the fact that each of them houses a protozoan which in turn houses a bacterium that possesses the cellulase enzyme. Enzymatic capabilities that were "left behind" in the bacterial world a billion years ago are not reinvented.

5. For a discussion of competing theories of the origins of life see Moss 1998.

6. Lynn Margulis is responsible for establishing the symbiotic basis of the evolution of the eukaryotic cell, among many other things (see Margulis 1981, 1996).

7. Individuals with no more than 29 CAG repeats do not appear to suffer disease symptoms. Most individual's who present with the disease have been seen to have between 40 and 55 repeats, with over 70 repeats being rare, but over 100 repeats occassionaly observed (see http://www.bcm.tmc.edu/neurol/struct/hunting/huntp3.html and related articles on the HD Clinic website for current findings).

8. For detailed discussion of this see Susan Oyama's classic *The Ontogeny of Information* (1985/2000) as well her numberous essays collected in *Evolution's Eye: A Systems View of the Biology-Culture Divide* (2000).

9. The term "teleonomic" was introduced by Pittendrigh (1958) to describe an "end-directed" process while not implying that a future state is an active agent in bringing its own realization. The use of teleonomic in place of teleological was championed especially by Ernst Mayr (see Mayr 1988).

References

Allen, C. and Bekoff, M. (1998), "Biological Function, Adaptaton, and Natural Design," in C. Allen, M. Bekoff, and G. Lauder (eds.), *Nature's Purposes Analysis of Function and Design in Biology*, Cambridge, MA: MIT Press.

Allen, G. (1975), *Life Science in the 20th Century*, New York: John Wiley and Sons.

Allen, G. (1985), "T. H. Morgan and the Split Between Embryology and Genetics, 1910–35" in T. J. Horder, J. A. Witkowski, and C. C. Wylie (eds.), *A History of Embryology*. Cambridge: Cambridge University Press.

Aristotle (1937), *Part of Animals* (translated by A. L. Peck), Cambridge MA: Harvard University Press.

Aristotle (1942), *Generation of Animals* (translated by A. L. Peck), Cambridge, MA: Harvard University Press.

Balme, D. M. (1972), *Aristotle's De Partibus Animalium I and De Generatione Animalium I*, Oxford: Clarendon Press.

Baltimore, D. (2001), "Our Genome Unveiled," *Nature*, **409**: 814–816.

Bell, G. et al. (1991), "Positive autoregulation of *Sex-lethal* by alternative splicing maintains the female determined state in *Drosophilia*," *Cell* **65**: 229–239.

Bennett (1978), "Purification of an active proteolytic fragment of the membrane attachment site for human erythrocyte spectrin," *Journal of Biological Chemistry* **253**: 2753.

Benjamin, W. (1969), "The Task of the Translator" in *Illuminations*, New York: Schocken.

Berezney, R. and Kwang, J. (eds.) (1995), *Nuclear Matrix: structural and functional organization*, San Diego: Academic Press.

Bishop, J. M. (1982), "Retroviruses and Cancer Genes," *Advances in Cancer Research* **37**: 1–32.

Bishop, J. M. (1991), "Molecular Themes in Oncogenesis," *Cell* **64**: 235–248.

Blau, H., Chiu, C., and Webster, C. (1983), "Cytoplasmic activation of human nuclear genes in stable heterocaryons," *Cell* **32**: 1171–1180.

Blau, H., Chiu, C., Pavlath, G., and Webster, C. (1985), "Muscle gene expression in heterokaryons" *Advances in Experimental Medicine Biology* **182**: 231–247.

Boland, C. R. et al. (1995), "Microallelotyping defines the sequence and tempo of allelic losses at tumour suppressor gene loci during colorectal cancer progression," *Nature Medicine* **1**: 902–909.

Boudreau, N. and Bissell, M. J. (1998), "Extracellular matrix signaling: Integration of form and function, in Normal and Malignant Cells," *Current Opinion in Cell Biology* **10**: 640–646.

Brandon, R. (1990), *Adaptation and Environment*, Princeton: Princeton University Press.

Brockes, J. P. (1998), "Regeneration and Cancer," *Biochem. et Biophys. Acta* **1377**: M1–M11.

Burdette, W. J. (1955), "The Significance of Mutation in Relation to the Origin of Tumors: A Review," *Cancer Research* **15**(4): 201–226.

Buss, L. W. (1987), *The Evolution of Individuality* Princeton: Princeton University Press.

Carlson, E. A. (1966), *The Gene: A Critical History*, Philadelphia: Saunders.

Cherry, R. J. (1979), "Rotational and lateral diffusion of membrane proteins," *Biochem. Biophys. Acta* **559**: 289.

Churchill, F. (1991), "The Rise of Classical Descriptive Embryology" in Scott Gilbert (ed.), *A Concise History of Modern Embryology*, Baltimore: Johns Hopkins University Press.

Cooper, G. M. (1990), *Oncogenes*, Boston: Jones and Bartlett.

Darden, L. (1991), *Theory Change in Science: Strategies from Mendelian Genetics*, New York: Oxford University Press.

Dawkins, R. (1976), *The Selfish Gene*, Oxford: Oxford University Press.

Dawkins, R. (1978), "Replicator Selection and the Extended Phenotype," *Zeitschrift Tierpsychol.*, 61–76.

Dawkins, R. (1982), *The Extended Phenotype*, Oxford: Oxford University Press.

Deng, G. et al. (1996), "Loss of heterozygosity in normal tissue adjacent to breast carcinomas," *Science* **274**: 2057–2059.

Dennett, D. C. (1995), *Darwin's Dangerous Idea*, New York: Simon and Schuster.

Depew, D. J. (1997), "Etiological approaches to biological aptness 1: Aristotle & Darwin" in W. Kollman, and S. Föllinger (eds.), Aristotelische Biologie: Intentionen, Methoden, Ergebnisse. Stuttgant: Franz Steiner Veriag.

Depew, D. and Weber, B. (1995), *Darwinism Evolving Systems Dynamics and the Genealogy of Natural Selection*, Cambridge, MA: MIT Press.

Depew, D. and Weber, B. (1998), "What does natural selection have to be like in order to work with self-organization?," *Cybernetics and Human Knowing*, 5: 18–31.

Depew, D. and Weber, B. (2001), "Developmental Systems, Darwinian Evolution, and the Unity of Science" in S. Oyama, P. Griffiths, and R. Gray (eds.), *Cycles of Contingency—Developmental Systems and Evolution*, Cambridge, MA: MIT Press.

Diamond, M. et al. (1990), "Transcriptional factor interactions: selectors of positive or negative regulation from a single DNA element," *Science* **249**: 1266–1272.

Dix, D., Cohn, P., and Flanery, J. (1980), "On the role of aging in cancer incidence," *Journal of Theoretical Biology* **83**: 163–173.

Doyle, R. (1997), *On Beyond Living: Rhetorical Transformations of the Life Sciences* Stanford: Stanford University Press.

Duesberg, P. (1983), "Retroviral Transforming Genes in Normal Cells?," *Nature* **304**: 219–226.

Dyson, F. (1999), *Origins of Life* (revised ed.), Cambridge: Cambridge University Press.

Elsasser, W. (1987), *Reflections on a Theory of Organisms*, Quebec: Editions Orbis Publishing.

Faber, K. (1893), "Multinucleate cells as phagocytes," *Journal of Pathological Bacteriology* **1**: 349.

Falk, R. (1991), "The Dominance of Traits in Genetic Analysis," *Journal of the History of Biology*, **24**: 457–484.

Falk, R. (1995), "The Struggle of Genetics for Independence," *Journal of the History of Biology* **28**: 219–246.

Farber, E. (1987), "Experimental induction of hepatocellular carcinoma as a paradigm for carcinogenesis," *Clin. Physiol. Biochem.* **5**: 152–159.

Farber, E. (1991), "Clonal adaptation as an important phase of hepatocarcinogenesis," *Cancer Biochem. Biophys.* **12**: 157–165.

Farley, J. (1977), *The Spontaneous Generation Controversy from Descartes to Oparin*. Baltimore: Johns Hopkins University.

Fearon, E. R. and Vogelstein, B. (1990), "A Genetic Model for Colorectal Tumorigenesis," *Cell* **61**: 759–767.

Fedoroff, N. (1999), "Transposable Elements as a Molecular Evolutionary Force," in L. Caporale (ed.), *Molecular Strategies in Biological Evolution, Annals of the New York Academy of Sciences* **870**: 251–264.

Forsberg, E. C. and Bresnick, E. H. (2001), Histone Acetylation Beyond Promoters. *Bioessays* **23**: 820–830.

Foulds, L. (1969), *Neoplastic Development*, New York: Academic Press.

Frankel, J. (1983), "What are the developmental underpinnings of evolutionary changes in protozoan evolution?," in B. C. Goodwin et al. (eds.), *Development and Evolution*, Cambridge: Cambridge University Press.

Frye, E. and Edidin, M. (1970), "The rapid intermixing of cell surface antigens after formation of heterokaryons," *Journal of Cell Science* 7: 319.

Gamow, G. (1954), "Possible Relation between Deoxyribonucleic Acid and Protein Structure," *Nature* 173: 318.

Gerhart, J. and Kirschner, M. (1997), *Cells, Embryos, and Evolution*, Malden, Mass: Blackwell Science.

Gilbert, S. (1978), "The Embryological Origins of the Gene Theory," *Journal of the History of Biology* 11: 307–351.

Godfrey-Smith, P. (2000), "The Replicator in Retrospect," *Biol. Philos.* 15: 403–423.

Goodman et al. (1983), "The spectrin membrane skeleton of normal and abnormal human erythrocytes: a review," *American Journal of Physiology* 244: C121.

Gottlieb, G. (1992), *Individual Development and Evolution: The Genesis of Novel Behavior*, New York: Oxford University Press.

Gray, Russel (1992), "Death of the Gene: Developmental Systems Strike Back," in P. Griffiths (ed.), *Trees of Life*, Dordrecht: Klewer.

Griesemer, J. (2000), "Reproduction and the Reduction of Genetics," in R. Beurton, R. Falk, and H.-J. Rheinberger (eds.), *The Concept of the Gene in Development and Evolution*, Cambridge: Cambridge University Press.

Griffiths, P. and Gray, R. (1994), "Developmental Systems and Evolutionary Explanation," *Journal of Philosophy* XCI: 277–304.

Griffiths, P. and Gray, R. (2001), "Darwinism and Developmental Systems," in S. Oyama, P. Griffiths, and R. Gray (eds.), *Cycles of Contingency—Developmental Systems and Evolution* Cambridge, MA: MIT Press.

Grimes, S. R. and Aufderheide, K. J. (1991), "Cellular aspects of pattern formation: the problem of assembly," *Monographs in Developmental Biology, vol 22*, Basel: Karger.

Haffner, R. and Oren, M. (1995), "Biochemical Properties and Biological Effects of p53," *Current Opinion in Genetics and Development* 5: 84–90.

Haggard, H. W. and Smith, G. M. (1938), "Johannes Müller and the Modern Conception of Cancer," *Yale Journal of Biology and Medicine* 10(5): 420–436.

Hall, B. K. (2000), "Evo-devo or devo-evo--does it matter?" *Evol. Devel.* 2: 177–179.

Hargreaves et al. (1980), "Reassociation of ankyrin with band 3 in erythrocyte membranes and in lipid vesciles," *Journal of Biological Chemistry* 255: 11965.

Harris, H. (1970), *Cell Fusion: The Dunham Lectures*, Cambridge, MA, Harvard University.

Harris, H. (1988), "The Analysis of Malignancy by Cell Fusion: The Position in 1988," *Cancer Research* 48: 3302–3306.

Harris, H. (1990), "The Role of Differentiation in the Suppression of Malignancy," *Journal of Cell Science* 97: 5–10.

Harris, H., Miller, O. J. et al. (1969), "Suppression of Malignancy by Cell Fusion," *Nature* **223**: 363–368.

Harris, H. and Watkins, J. F. (1965), "Hybrid Cells Derived from Mouse and Man: Artificial Heterokaryons of Mammalian Cell from Different Species," *Nature, London* **205**: 640.

Harwood, J. (1993), *Styles of Scientific Thought: The German Genetics Community 1900–1933*, Chicago: University of Chicago Press.

Heidegger, M. (1977), "The Age of the World Picture" in *The Question Concerning Technology and Other Essays*, New York: Harper Colophon Books.

Hodge, M. J. S. (1985), "Darwin as a Lifeling Generation Theorist," in *The Darwinian Heritage*, in D. Kohn (ed.), Princeton: Princeton University Press.

Huebner, R. J. and Todero, G. J. (1969), "Oncogenes of RNA Tumor Viruses as a Determinant of Cancer," *Proceedings of the National Academy of Science USA* **64**: 1087–1094.

Hull, D. (1980), "Individuality and Selection," *Ann. Rev. Ecol. Systematics* **11**: 311–332.

Hulsken, J., Behrens, J. et al. (1994), "Tumor-Suppressor Gene Products in Cell Contacts: the Cadherin-APC-Armadillo Connection," *Current Opinion in Cell Biology* **6**(5): 711–716.

Hunter, T. (1995), "Protein kinases and phosphotases: The yin and yang of protein phosphorylation and signalling," *Cell* **80**: 225–236.

Huxley, J. (1958), *Biological Aspects of Cancer*, London: George Allen and Unwin.

Jablonka, E. (2001), "The Systems of Inheiritance" in S. Oyama, P. Griffiths, and R. Gray (eds.), *Cycles of Contingency—Developmental Systems and Evolution*, Cambridge, MA: MIT Press.

Jablonka, E. and Lamb, M. J. (1995), *Epigenetic Inheritance and Evolution*, Oxford: Oxford University Press.

Jennwein, T. and Allis, C. D. (2001), "Translating the Histone Code," *Science* **293**: 1074–1080.

Jilling, T. and Kirk, K. (1997), "The Biogenesis, Traffic, and Function of the Cystic Fibrosis Transmembrane Conductance Regulator," *International Review of Cytology* **172**: 193–241.

Johannsen, W. (1911), "The Genotype Conception of Heredity," *The American Naturalist* **45**: 129–159.

Johannsen, W. (1923), "Some Remarks About Units in Heredity," *Hereditas* **4**: 133–141.

Judson, H. F. (1979), *The Eight Day of Creation*, New York: Simon and Schuster.

Kaneko, K. and Yomo, T. (1994), "A Theory of Differentiation with Dynamic Clustering," *Physica D* **75**: 89–102.

Kant, I. (1987), *The Critique of Judgment*, Indianapolis: Hackett.

Kauffman, S. (1993), *The Origins of Order*, Oxford: Oxford University Press.

Kauffman, S. (1995), "What is Life?: Was Schrödigner Right?" in M. P. Murphy and L. A. J. O'Neill (eds.), *What is Life? The Next Fifty Years*, Cambridge: Cambridge University Press.

Kay, L. (2000), *Who Wrote the Book of Life? A History of the Genetic Code*, Stanford: Stanford University Press.

Keller, E. F. (2001), *The Century of the Gene*, Cambridge, MA: Harvard University Press.

Kelly, D. E., Wood, R. L. et al. (1984), *Bailey's Textbook of Microscopic Anatomy*. Baltimore and London: Williams and Wilkins.

Kerem, E. and Kerem, B. (1995), "The relationship between genotype and phenotype in cystic fibrosis," *Current Opinion in Pulmonary Medicine* 1: 450–456.

Kinzler, K. W. and Vogelstein, B. (1996), "Lessons from hereditary colorectal cancer," *Cell* 87: 159–170.

Kirschner, M. and Gerhart, J. (1998), "Evolvability," *Proc. Natl. Acad. Sci.* 95: 8420–8427.

Kuhn, T. (1962), *The Structure of Scientific Revolutions*, Chicago: University of Chicago Press.

Lander, E. et al. (International Human Genome Sequencing Consortium) (2001), "Initial Sequencing and Analysis of the Human Genome," *Nature*, 409: 860–921.

Lawley, P. D. (1994), "Historical Origins of Current Concepts of Carcinogenesis," *Advances in Cancer Research* 65: 17–111.

Lenoir, T. (1980), "Kant, Blumenbach, and Vital Materialism in German Biology," *Isis* 71(256): 77–108.

Lenoir, T. (1982), *The Strategy of Life: Teleology and Mechanics in 19th Century German Biology*, Chicago: University of Chicago Press.

Levine, A. J. (1993), "The Tumor Suppressor Genes," *Annual Review of Biochemistry* 62: 623–651.

Luna, E. J. et al. (1979), "Identification by peptide analysis of the spectrin binding protein in human erythrocyte membranes," *Journal Cell Biology* 59: 395.

Maienschein, J. (1987), *Defining Biology: Lectures from the 1890s*, Cambridge MA: Harvard University Press.

Marshall, C. J. (1991), "Tumor Suppressor Genes," *Cell* 64: 313–326.

Maturana, H. and Varela, F. (1973), *Autopoiesis The Organization of the Living. Boston Studies in the Philosophy of Science*, vol. 42, Dordrecht: D. Reidel Publishing Co.

Matzke, M. A., Mette, M. F., and Matzke, A. T. M. (2000), "Transgene Silencing by the Host Genome Defense: Implications for the Evolution of Epigenetic

Control Mechanisms in Plants and Vertebrates," *Plant Molecular Biology* 43: 401–415.

Mayr, E. (1988), *Towards a New Philosophy of Biology*, Cambridge, MA: Harvard University Press.

McCulloagh, K. D. et al. (1997), "Age-dependent induction of hepatic tumor regression by the tissue microenvironment after transplantation of neoplastically transformed rat liver epithelial cells into the liver," *Cancer Research* 57: 1807–1813.

Miller, S., Pavlath, G., Blakely, B., and Blau, H. (1988), "Muscle cell components dictate hepatocye gene expression and the distribution of the Golgi apparatus in heterokaryons," *Gene Development* 2: 330–340.

Mintz, B., Cronmiller, C., and Coster R. P. (1978), "Somatic cell origin of teratocarcinomas," *Proc. Natl. Acad. Sci.* 75: 2834–2838.

Mondal, S. and Heidelberger, C. (1970), "In vitro malignant transformation by methylcholanthrene of the progeny of single cells derived from C3H mouse prostate," *Proc. Natl. Acad. Sci.* 65: 219–225.

Morgan, T. (1910), "Chromosomes and Heredity," *The American Naturalist* 44: 449–496.

Moss, L. (1992), "A Kernal of Truth? On the Reality of the Genetic Program," in D. Hull, A. Fine, and M. Forbes (eds.), *Proceedings of the Philosophy of Science Association 1992* vol. 1: 335–348. East Lansing, MI: The Philosophy of Science Association.

Moss, L. (1998), "Life, Origins of," in E. Craig (ed.), *Routledge Encyclopedia of Philosophy*, London: Routledge.

Moss, L. (2001), "Deconstructing the Gene and Reconstructing Molecular Developmental Systems" in S. Oyama, P. Griffiths, and R. Gray (eds.), *Cycles of Contingency—Developmental Systems and Evolution*, Cambridge, MA: MIT Press.

Nigg, E. A. and Cherry, R. J. (1980), "Anchorage of a band 3 population at the erythrocyte cytoplasmic membrane surface: protein rotational diffusion measurments," *Proceedings of the National Academy of Sciences* 77: 4702.

Novick, A. and Weiner, M. (1957), "Enzyme induction as an all-or-none phenomenon," *Proc. Natl. Acad. Sci.* 43: 533–566.

Olby, R. (1974), *The Path to the Double Helix*, New York: Dover.

Osler, Margaret J. (1994), *Divine Will and the Mechanical Philosophy: Gassendi and Descartes on Contingency and Necessity in the Created World*, Cambridge: Cambridge University Press.

Oyama, S. (1985/2000), *The Ontogeny of Information* (revised edition), Durham, N.C.: Duke University Press.

Oyama, S. (2000), *Evolution's Eye: A Systems View of the Biology-Culture Divide*, Durham: Duke University Press.

Oyama, S., Griffiths, P., and Gray, R. (eds.) (2001), *Cycles of Contingency—Developmental Systems and Evolution*, Cambridge MA: MIT Press.

Park, C., Bissell, M. J., and Barcellos-Hoff, M. H. (2000), "The Influence of the Microenvironment on the Malignant Phenotype," *Molecular Medicine Today* **6**: 324–329.

Pittendrigh, C. (1958), "Adaptation, Natural Selection, and Behavior," in A. Roe and G. G. Simpson (eds.), *Behavior and Evolution*, New Haven: Yale University Press.

Poo, C. (1974), "Lateral diffusion of rhodopsin in the photoreceptor membrane," *Nature* **247**: 438.

Portugal, F. H. and Cohen, J. S. (1977), *A Century of DNA*, Cambridge: Cambridge University Press.

Prigogine, I. (1980), *From Being to Becoming: Time and Complexity in the Physical Sciences*, San Francisco: W. H. Freeman.

Radisky, D., Hagios, C., and Bissell, M. J. (2001), "Tumors are Unique Organs Defined by Abnormal Signaling and Context," *Seminars in Cancer Biology* **11**: 87–95.

Radman, M. (1973), "Phenomenology of an Inducible Mutagenic DNA Repair Pathway in *Escherichia coli*: SOS Repair Hypothesis," in L. Prakash et al. (eds.), *Molecular and Environmental Aspects of Mutagenesis*, Springfield, IL: Charles C. Thomas.

Radman, M. (1999), "Mutation: Enzymes of Evolutionary Change," *Nature*, **401**: 866–869.

Rather, L. J. (1978), *The Genesis of Cancer*, Baltimore: Johns Hopkins University Press.

Rather, L. J. and Rather, P. et al. (1986), *Johannes Müller and the 19th-Century Origins of Tumor Cell Theory*, Canton, MA: Science History Publications.

Rice, J. C. and Allis, C. D. (2001), "Histone Methylation Versus Histone Acetylation: New Insights Into Epigenetic Regulation," *Current Opinion in Cell Biology* **13**: 263–273.

Richards, R. J. (1992), *The Meaning of Evolution*, Chicago: University of Chicago Press.

Richards, R. J. (2000), "Kant and Blumenbach on the *Bildungstrieb*: A Historical Misunderstanding," *Stud. Hist. Phil. Biol. & Biomed. Sci.* **31**: 11–32.

Roe, S. A. (1981), *Matter, Life, and Generation: Eighteenth-century embryology and the Haller-Wolff debate*, Cambridge: Cambridge University Press.

Rothman, J. E. and Wieland, F. T. (1996), "Protein Sorting by Transport Vesicles," *Science* **272**: 227–234.

Rous, P. (1959), "Surmise and Fact on the Nature of Cancer," *Nature* **183**: 1357–1361.

Rubin, H. (1955), "Quantitative relations between causative virus and cell in the Rous no. 1. Chicken Sarcoma." *Proc. Natl. Acad. Sci. USA* **1**: 445–473.

Rubin, H. (1965), "The nucleic acid of the Bryan strain of Rous sarcoma virus: purification of the virus and isolation of the nucleic acid," *Proc. Natl. Acad. Sci. USA* **54**: 137–144.

Rubin, H. (1993), " 'Spontaneous' transformation as aberrant epigenesis," *Differentiation* **53**: 123–137.

Rubin, H., Yao, A., and Chow, M. (1995a), "Neoplastic development: Paradoxical relations between impaired cell growth at low population density and excessive growth at high density," *Proc. Natl. Acad. Sci. USA* **92**: 7734–7738.

Rubin, H., Yao, A., and Chow, M. (1995b), "Heritable, population-wide damage to cells as the driving force of neoplastic transformation," *Proc. Natl. Acad. Sci. USA* **92**: 4843–4847.

Rubin, H. (1997), "Cell Aging in vivo and in vitro," *Mechanisms of Aging and Development* **98**: 1–35.

Rubin, H. (1999), "Neoplastic Transformation," in T. Creighton (ed.), *Encyclopedia of Molecular Biology* New York: John Wiley & Sons.

Saffman, D. (1975), "Brownian Motion in Biological Membranes" *Proceedings of the National Academy of Sciences* **72**: 3111.

Sapp, J. (1987), *Beyond the Gene: Cytoplasmic Inheritance and the Struggle for Authority in Genetics*, Oxford: Oxford University Press.

Sarkar, S. (1999), "From Reaktionsnorm to the Adaptive Norm: The Norm of Reaction, 1909–1960," *Biology and Philosophy* **14**: 235–252.

Sasaki, J. and Yoshida, T. (1935), *Virchows Arch. F. Path. Anat.* **295**: 175–200.

Schlessinger, J. et al. (1978), "Quantitative determination ofthe lateral diffusion coefficients of the hormone-receptor complexes of insulin and epidermal growth factor onthe plasma membrane of cultured fibroblasts," *Proceedings of the National Academy Sciences* **75**: 5353.

Schlichting, C. D. and Pigliucci, M. (1998), *Phenotypic Evolution: A Reaction Norm Perspective*, Sunderland, Mass: Sinauer Associates.

Schrödinger, E. (1944), *What is Life?*, Cambridge: Cambridge University Press.

Schwab, M. (1998), "Amplification of Oncogenes in Human Cancer," *BioEssays* **20**: 473–479.

Shapiro, J. A. (1995), "The Significances of bacterial colony patterns," *BioEssays* **17**: 597–607.

Shapiro, J. A. (1999), "Genome System Architecture and Natural Genetic Engineering in Evolution" in Caporale (ed.), *Molecular Strategies in Biological Evolution, Annals of the New York Academy of Sciences* **870**: 23–25.

Singer, J. and Nicolson, G. (1972), "The Fluid Mosaic Model of the Structures of Cell Membranes," *Science* **175**: 720.

Sloan, P. (2001), "Preforming the Categories: 18th C. Generation Theory and the Biological Roots of Kant's A-Priori," *Journal of the History of Philosophy*.

Smith, B. A. et al. (1979), "Anisotropic molecular motion on cell surface," *Proceedings of the National Academy Sciences* 76: 5641.

Smithers, D. W. (1962), "Cancer: An Attack on Cytologism" *Lancet*, no. 7228 (10 March 1962): 493–499.

Sonnenschein, C. and Soto, A. M. (1999), *The Society of Cells: Cancer and Control of Cell Proliferation*, New York: Springer-Verlag.

Stanbridge, E. J. (1990), "Human Tumor Suppressor Genes," *Annual Review of Genetics* 24: 615–657.

Stent, G. (1995), "The Aperiodic Crystal of Heredity," *Annals of the New York Academy of Sciences* 758: 25–31.

Sterelny, K., Smith, K., and Dickison, M. (1996), "The Extended Replicator," *Biol. Philos.*, 11: 377–403.

Stern, C. and Sherwood, E. (1966), *The Origins of Genetics. A Mendel Sourcebook*, San Francisco: Freeman.

Summerhayes, I. C. and Franks, L. M. (1979), "Effects of donor age on neoplastic transformation of adult mouse bladder epithelium in vitro," *J. Natl. Cancer Inst.* 62: 1017–1023.

Symer, D. and Bender, J. (2001), "Hip-hopping Out of Control," *Nature*, 411: 146–148.

Temin, H. M. (1971), "The Protovirus Hypothesis: Speculations on the Significance of RNA Directed DNA Synthesis for Normal Development and for Carcinogenesis," *Journal of the National Cancer Institute* 46: iii–vii.

Triolo, V. A. (1964), "Nineteenth Century Foundations of Cancer Research—Origins of Experimental Research," *Cancer Research* 24: 4–27.

Triolo, V. A. (1965), "Nineteenth Century Foundations of Cancer Research: Advances in Tumor Pathology, Nomenclature, and Theories of Oncogenesis," *Cancer Research* 25(2): 75–106.

Tupler, R., Perini, G., and Green, M. (2001), "Expressing the Human Genome," *Nature* 409: 832–833.

Varmus, H. (1989), "Nobel Lecture—Retroviruses and Oncogenes," *Bioscience Reports* 10: 413–429.

Vaz et al. (1981), "Translational Mobility of Glycophorin in Bilayer Membranes of Dimyristoylphosphotidyl Choline," *Biochemistry* 20: 1392.

Venter, J. C. et al. (2001), "The Sequence of the Human Genome," *Science* 291: 1304–1351.

Volosinov, V. N. (1973), *Marxism and the Philosophy of Language*, Cambridge, MA: Harvard University Press.

De Vries (1900), "Das Spaltungsgestz der Bastarde," *Berichte der deutschen botanischen Gesellschaft* 18: 83–90, reprinted in English translation as "The Law of Segregation of Hybrids," in C. Stern, and E. Sherwood (eds.) (1966), *The Origin of Genetics: A Mendel Source Book*, San Francisco: W. H. Freeman and Co.

Walsh, F. and Doherty, P. (1997), "Neural cell adhesion molecules of the immunoglobulin superfamily: Role in axon growth and guidance," *Annual Review of Cell and Develomental Biology* 13: 425–456.

Weber, B. and Depew, D. (1996), "Natural Selection and Self-Organization: Dynamical Models as Clues to a New Evolutionary Synthesis," *Biology and Philosophy* 11: 33–65.

Weinberg, R. A. (1989), "*Multistep Carcinogenesis,*" in R. A. Weinberg (ed.), *Oncogenes and the Molecular Origins of Cancer*, Cold Spring Harbor, New York, Cold Spring Harbor Laboratory.

Weiss, Paul (1940), "The Problem of Cell Individuality in Development," *American Naturalist* 74: 34–46.

Werb, Z., Sympson, C. J., Alexander, C. M., Thomasset, N., Lund, L. R., Mac Auleg, A., Ashkenas, V., and Bissell, M. J. (1996), "Extracellular Matrix Remodeling and the Regulations of epithelial-stromal Interactions during differentiation and involution," *Kidney International* Suppl. 54: 568–574.

Whitman, C. O. (1893), "The Inadaquacy of the Cell-Theory of Development," *Journal of Morphology* 8: 639–658.

Wilson, E. B. (1893), "The Mosaic Theory of Develoment" in J. Maienschein (ed.) (1987), *Defining Biology: Lectures from the 1890s*, Cambridge, MA: Harvard University Press.

Wimsatt, W. C. (1986), "Developmental Constraints, Generative Entrenchment, and the Innate-acquired Distinction", in W. Bechtel (ed.), *Integrating Scientific Disciplines*, The Hague: Martinus-Nijhoff.

Wimsatt, W. C. (1999), "Generativity, Entrenchment, Evolution, and Innateness," in V. Hardcastle (ed.), *Biology Meets Psychology*, Cambridge, MA: MIT Press (forthcoming).

Wolf, U. (1995), *Human Genetics* 95: 127–148.

Wright, S. (1945), "Genes as Physiological Agents. General Considerations," *American Naturalist* 74: 289–303.

Wu, J. T. et al. (1981), "Lateral Diffusion of Concanavalin A Receptors and Lipoid Analogs in Normal and Bulbous Lymphocytes," *Biophysics Journal* 33: 74a.

Wyllie, A. H. (1995), "The Genetic Regulation of Apoptosis," *Current Opinion in Genetics and Development* 5: 97–104.

Zorn, A. and Krieg, P. (1992), "Developmental Regulation of Alternative Splicing in the mRNA Encoding *Xenopus laevis* Neural Cell Adhesion Molecule (NCAM)," *Developmental Biology* 149: 197–205.

Zimmerman, J. A. et al. (1982), "Pancreatic Carcinoma Induced by N-methyl-N'-nitrosourea in Aged Mice," *Gerontology* 28: 114–120.

Index